WILDERNESS LAWS: PROVISIONS, PERMISSIONS AND PROHIBITIONS

WILDERNESS LAWS: PROVISIONS, PERMISSIONS AND PROHIBITIONS

Cooper Brown
and
Thomas A. Wilson
EDITORS

Nova Science Publishers, Inc.
New York

For permission to use material from this book please contact us:
Telephone 631-231-7269; Fax 631-231-8175
Web Site: http://www.novapublishers.com

Library of Congress Cataloging-in-Publication Data

Gorte, Ross W.
Wilderness laws : provisions, permissions and prohibitions / [Ross W. Gorte] ; editors, Cooper Brown and Thomas A. Wilson.
p. cm.
"CRS report for Congress prepared for members and committees of Congress."
Includes bibliographical references and index.
ISBN 978-1-61470-694-6 (hardcover : alk. paper) 1. Wilderness areas--Law and legislation--United States. 2. Forest reserves--Law and legislation--United States. I. Brown, Cooper. II. Wilson, Thomas A., 1972- III. Library of Congress. Congressional Research Service. IV. Title.
KF5635. A25 2011
346.7304'6782--dc23
2011025578

Published by Nova Science Publishers, Inc. † New York

CONTENTS

PREFACE

Congress enacted the Wilderness Act in 1964. It established a National Wilderness Preservation System of federal lands "where the earth and its community of life are untrammeled by man, where man himself is a visitor who does not remain." The act designated 54 wilderness areas containing 9.1 million acres of federal land within the national forests. It also reserved to Congress the authority to add areas to the system, although it also directed agencies to review the wilderness potential of certain lands. This book summarizes the various statutory provisions and provisions on prohibited and permitted uses within wilderness areas.

Chapter 1- The 1964 Wilderness Act established a National Wilderness Preservation System of federal lands "where the earth and its community of life are untrammeled by man, where man himself is a visitor who does not remain." The act designated 54 wilderness areas with 9.1 million acres within the national forests and reserved to Congress the authority to add areas to the system. Congress has enacted 117 subsequent statutes designating wilderness areas (including one with 16 wilderness-related subtitles) and 8 other statutes requiring wilderness study or otherwise significantly affecting wilderness areas. Many of these statutes provide management direction for designated areas that differs from the Wilderness Act provisions. As of December 31, 2010, the system totaled 759 wilderness areas with 109.7 million acres of federal land.

Chapter 2- Congress enacted the Wilderness Act in 1964. This act created the National Wilderness Preservation System, reserved to Congress the authority to designate wilderness areas, and directed the Secretaries of Agriculture and of the Interior to review certain lands for their wilderness potential. The act also designated 54 wilderness areas with 9 million acres of federal land. Congress began expanding the Wilderness System in 1968, and today, there are 759 wilderness areas, totaling nearly 110 million acres, in 44 states. Numerous bills to designate additional areas and to expand existing ones are introduced and considered in every Congress.

In: Wilderness Laws: Provisions, Permissions ... ISBN: 978-1-61470-694-6
Editors: C. Brown, T. A. Wilson

Chapter 1

WILDERNESS LAWS: STATUTORY PROVISIONS AND PROHIBITED AND PERMITTED USES*

Ross W. Gorte

SUMMARY

The 1964 Wilderness Act established a National Wilderness Preservation System of federal lands "where the earth and its community of life are untrammeled by man, where man himself is a visitor who does not remain." The act designated 54 wilderness areas with 9.1 million acres within the national forests and reserved to Congress the authority to add areas to the system. Congress has enacted 117 subsequent statutes designating wilderness areas (including one with 16 wilderness-related subtitles) and 8 other statutes requiring wilderness study or otherwise significantly affecting wilderness areas. Many of these statutes provide management direction for designated areas that differs from the Wilderness Act provisions. As of December 31, 2010, the system totaled 759 wilderness areas with 109.7 million acres of federal land.

The Wilderness Act and other wilderness statutes have contained many provisions related to the administration of the areas. All but three

---

* This is an edited, reformatted and augmented version of a Congressional Research Service publication, CRS Report for Congress R41649, from www.crs.gov, dated February 22, 2011.

direct *management in accordance with the Wilderness Act.* Provisions prohibiting *buffer zones* around designated areas are common. Many also preserve existing *state jurisdiction and responsibilities over fish and wildlife*, while some preserve *other jurisdictions and authorities*, such as for law enforcement and cooperation with other federal, state, and local agencies. *Water rights* has been a controversial issue—some statutes have neither claimed nor denied water rights, some have reserved water rights, and others have directed no claim to water. Several statutes have directed *wilderness study* of potentially qualified lands, and have designated *intended or potential wilderness*, contingent upon some future condition or event. Concern about protection of the study areas has led Congress to include provisions addressing *interim management and release* of areas during and after the studies.

The Wilderness Act generally prohibits commercial activities within wilderness areas, although it allows commercial activities related to wilderness-type recreation. The act also generally prohibits motorized and mechanical access, and roads, structures, and other facilities within wilderness areas. Although wilderness is generally open to other public uses, some wilderness statutes have authorized temporary closures for various reasons. Also, many statutes have withdrawn the designated areas from the public land disposal laws, the mining and mineral leasing laws, and from the laws authorizing the disposal of common mineral materials. However, valid existing rights are not terminated, and can be developed under reasonable regulations.

The Wilderness Act and many subsequent wilderness statutes have also allowed various nonconforming uses and conditions. Motorized access has generally been permitted for *management requirements and emergencies*, for *nonfederal inholdings*, and for *fire, insect, and disease control. Continued motorized access* and *livestock grazing* have also generally been permitted where they had been occurring prior to the area's designation as wilderness. Construction, operations, and maintenance, and associated motorized access, have also been permitted for *water infrastructure* and for *other infrastructure* in many instances. Motorized access for state agencies for *fish and wildlife management activities* has sometimes been explicitly allowed. *Low-level military overflights* of wilderness areas have been permitted in several statutes. Access for *minerals activities* has been authorized in some specific areas and for valid existing rights; the Wilderness Act specifically allowed for mineral prospecting and for establishing mineral rights for 20 years after enactment. Finally, several statutes have allowed access for *other specific activities*, such as access to cemeteries within designated areas or for tribal activities.

INTRODUCTION

Congress enacted the Wilderness Act (P.L. 88-577; 16 U.S.C. §§ 1131-1136) in 1964. It established a National Wilderness Preservation System of federal lands "where the earth and its community of life are untrammeled by man, where man himself is a visitor who does not remain." The act designated 54 wilderness areas containing 9.1 million acres of federal land within the national forests. It also reserved to Congress the authority to add areas to the system, although it also directed agencies to review the wilderness potential of certain lands.

The Wilderness Act and the 132 subsequent laws[1] designating wilderness contain numerous statutory provisions addressing management of wilderness areas as well as many provisions addressing prohibited and permitted uses, both generally and in specific areas.[2] This report summarizes the various statutory provisions and the provisions on prohibited and permitted uses within wilderness areas. Appendix A is a list of provisions in each relevant law discussed in the sections below. Appendix B includes a complete chronological list of laws designating wilderness areas, with a summary of or quotation from all the wilderness-related provisions in each law.[3] As of December 31, 2010, the National Wilderness Preservation System totaled 759 areas, with 109.7 million acres.[4] The wilderness areas are part of and within the existing units of federal land administered by the several federal land management agencies—the Forest Service (USFS) in the Department of Agriculture, and the National Park Service (NPS), Fish and Wildlife Service (FWS), and Bureau of Land Management (BLM) within the Department of the Interior.

The subsequent wilderness statutes have not designated wilderness areas by amending the Wilderness Act. Instead, they are independent statutes. While nearly all direct management in accordance with the Wilderness Act, as discussed below, most also provide unique management guidance for the areas designated in that statute. Thus, altering management direction for the entire National Wilderness Preservation System, for example to modify land acquisition authority, might require amending all the wilderness statutes, not just the Wilderness Act.

STATUTORY PROVISIONS

The Wilderness Act and subsequent wilderness laws contain several provisions addressing management of wilderness areas. These laws designate wilderness areas as part of and within existing units of federal land, and the management provisions applicable to those units of federal land, particularly those governing management direction and restricting activities, also apply. For example, hunting is prohibited in many NPS units, but not on USFS or BLM lands, and thus would be prohibited in wilderness areas in those NPS units but generally not in USFS or BLM wilderness areas, absent specific language.

Manage in Accordance with the Wilderness Act

The Wilderness Act identified the purposes of wilderness in § 2. Specifically, § 2(a) stated that the purpose was to create a National Wilderness Preservation System of federal lands

> administered for the use and enjoyment of the American people in such manner as will leave them unimpaired for future use and enjoyment as wilderness, and so as to provide for the protection of these areas, the preservation of their wilderness character, and for the gathering and dissemination of information regarding their use and enjoyment as wilderness...

The act goes on to further define wilderness area management in § 2(c):

> A wilderness, in contrast to those areas where man and his own works dominate the landscape, is hereby recognized as an area where the earth and its community of life are untrammeled by man, where man himself is a visitor who does not remain. An area of wilderness is further defined to mean in this Act an area of undeveloped Federal land retaining its primeval character and influence, without permanent improvements or human habitation, which is protected and managed so as to preserve its natural conditions and which (1) generally appears to have been affected primarily by the forces of nature, with the imprint of man's work substantially unnoticeable; (2) has outstanding opportunities for solitude or a primitive and unconfined type of recreation; (3) has at least five thousand acres of land or is of sufficient size as to make practicable its preservation and use in an unimpaired condition; and (4) may also

> contain ecological, geological, or other features of scientific, educational, scenic, or historical value.

All but three of the subsequent wilderness statutes—P.L. 90-532, P.L. 90-544, and P.L. 92-476— direct management of the designated areas in accordance with or consistent with the Wilderness Act. Thus, virtually all areas within the National Wilderness Preservation System must be managed under the purposes described above and under the various management directions included in the Wilderness Act, as described below. In addition, four statutes require management plans for the designated wilderness areas. For all other designated areas, management must be included in management plans for the unit or area which encompasses the designated wilderness.

Buffer Zones

The Wilderness Act is silent on the issue of buffer zones around wilderness areas to protect the designated areas. However, language in subsequent wilderness bills has prohibited buffer zones restricting uses and activities on federal lands around the wilderness areas. The first explicit language was enacted in 1980 in P.L. 96-550; § 105 states:

> Congress does not intend that the designation of wilderness areas ... lead to the creation of protective perimeters or buffer zones around each wilderness area. The fact that nonwilderness activities or uses can be seen or heard from areas within the wilderness shall not, of itself, preclude such activities or uses up to the boundary of the wilderness area.

Virtually identical language has been included in 30 other wilderness statutes enacted since 1980.

State Fish and Wildlife Jurisdiction and Responsibilities

The Wilderness Act explicitly directed that the wilderness designations had no effect on state jurisdiction or responsibilities over fish and wildlife; § 4(d)(8) states that "nothing in this Act shall be construed as affecting the jurisdiction or responsibilities of the several States with respect to wildlife and fish in the national forests."[5] Comparable language, sometimes only referring to state jurisdiction (not responsibilities), has been included in 31 wilderness statutes, beginning in 1978. Such provisions seem to be more common in recent legislation; for example, 8 of the 16 wilderness subtitles of P.L. 111-11 and five of the six wilderness statutes enacted in the 109th Congress included such language. Concern over state wildlife and fish management in wilderness areas persists, and several statutes have included additional specific provisions over permissible access and activities for fish and wildlife management; these are discussed under "Nonconforming Permitted Uses," below.

Jurisdiction and Authorities of Other Agencies

Several wilderness statutes have directed that other agencies' specific authorities, jurisdiction, and related activities be allowed to continue. Three—P.L. 101-628 (AZ), P.L. 103-433 (CA), and P.L. 106-145 (CA)—directed no effect on U.S.-Mexico border operations. P.L. 103-433 added no effect on law enforcement generally, and allowed motorized access for law enforcement and border operations. P.L. 106-145 added no effect on drug interdiction, and allowed motorized access subject to conditions established by the Secretary. Two other laws—P.L. 106-65 (AZ) and P.L. 111-11, Subtitle K (NM)—directed no effect on military training, for current and future aviation training and for an adjacent training center respectively. In addition, P.L. 95-495 directed cooperation with the State of Minnesota generally, while P.L. 98-550 directed cooperation with the State of Wyoming on cultural resource management. P.L. 107-282 directed no effect on Park Service management of the Lake Mead National Recreation Area. Finally, P.L. 111-11, Subtitle L, directed no effect on management of existing utilities outside the designated wilderness areas.

Land and Rights Acquisition

The Wilderness Act authorizes the acquisition of land within designated wilderness areas (called inholdings). Section 5(c) authorizes acquisition, subject to appropriations, "if (1) the owner concurs in such acquisition or (2) the acquisition is specifically authorized by Congress." In addition, § 5(a) authorizes acquisition of inholdings by exchange for other federal land of approximately equal value, but the exchange can grant mineral interests only if the landowner relinquishes mineral interests in the inholding. Section 6(a) authorizes the acceptance of gifts or bequests of land within or adjacent to the wilderness, and after 60 days notice to Congress shall become part of the designated wilderness.

Several subsequent wilderness statutes have provided specific directions on acquisitions within the areas designated in those statutes.[6] P.L. 93-622, the Eastern Wilderness Act, authorized acquisition through condemnation, as well as by purchase, gift, or exchange. P.L. 97-466 (WV) directed the acquisition of coal and other mineral interests, with detailed provisions on the valuation procedures and the use of credits for other federal mineral rights elsewhere; P.L. 104- 333 authorized the acquisition of mineral leases by exchange. P.L. 98-425 (CA) directed

negotiations for acquisition via an exchange, and P.L. 98-574 (TX) directed an expeditious land exchange with a forest products company. P.L. 100-184 (MI) explicitly required concurrence of the landowner for land acquisition. P.L. 101-628 (AZ) directed the acquisition of mineral rights by exchange, and P.L. 103-77 (CO) directed mineral right acquisition only by exchange or donation.

Water Rights

In contrast to the preceding statutory provisions, where Congress has been relatively consistent in the language used or has been silent on the particular issue, wilderness statutes have provided different directions concerning federal reserved water rights associated with the designated wilderness areas. Under the *Winters* doctrine, when Congress reserves federal land for a particular purpose, it also reserves enough water to fulfill the purpose of the reservation.[7] Congress also has repeatedly deferred to state law in the regulation of water allocation and use.[8]

State Authorities and Water Agreements

Numerous wilderness statutes direct that they are to have no effect on various water agreements and state jurisdiction over water rights. The first was P.L. 95-495, the Boundary Waters Canoe Area Wilderness Act, which directed no effect on Minnesota's jurisdiction or responsibilities over water rights and management. P.L. 96-550 directed no effect on management of a particular municipal watershed. Five statutes—P.L. 107-282 (NV), P.L. 108-424 (NV), P.L. 109-94 (CA), P.L. 109-432 (NV), and P.L. 111-11, Subtitle O (UT)—direct no effect on state water jurisdiction. These five statutes, plus P.L. 103-77 (CO) and P.L. 106-353 (CO), also direct that the statutes are not to be "construed as limiting, altering, modifying, or amending any interstate compacts or equitable apportionment decrees that apportion water among and between" the states. In addition, these five statutes, plus P.L. 111-11, Title II, Subtitle E (NM), direct that any water rights be secured under state law. Two statutes—P.L. 101-628 (AZ) and P.L. 104-433 (CA)—direct no effect on state, interstate, federal, or international jurisdiction, agreements, or treaties pertaining to the Colorado River. Finally, P.L. 108-447 (WI) directs the preservation of existing treaty rights and management of Lake Superior waters.

Neither Claim Nor Denial of Claim

The Wilderness Act, in § 4(d)(7), states that "nothing in this Act shall constitute an express or implied claim or denial on the part of the Federal Government as to exemption from State water laws." Comparable language—neither claiming nor denying reserved water rights for the wilderness designations outside of the state legal system for allocating water—has been used in five subsequent wilderness statutes—P.L. 96-312 (ID), P.L. 98-406 (AZ), P.L. 98-428 (UT), P.L. 98-550 (WY), and P.L. 106-399 (OR).

Reserved Water Rights

In contrast to the Wilderness Act, several subsequent wilderness statutes have expressly reserved federal water rights associated with the designated wilderness areas. Statutes with such an express reservation include P.L. 100-668 (WA), P.L. 101-195 (NV), P.L. 101-628 (AZ), P.L. 102-301 (CA), P.L. 103-433 (CA), and P.L. 107-370 (CA). Another four wilderness statutes (P.L. 107-282 (NV), P.L. 108-424 (NV), P.L. 109-94 (CA), and P.L. 109-432 (NV)) indirectly protect wilderness water flows by prohibiting federal funds, assistance, authorization, or permits for new water resource projects or facilities within the wilderness areas (except "water guzzlers" for wildlife in P.L. 109-94).

No Claim on Water

Also in contrast to the Wilderness Act and to the statutes identified above, several wilderness statutes have explicitly denied claims to water associated with the designated wilderness areas. As discussed below, this denial of water rights has taken two different forms, each in several statutes: the direction to have no effects on water rights in specific geographic areas; and the denial of a reserved water right for all the areas designated in the statute.

Area-Specific Provisions

Several wilderness statutes have specified that they are not to have any effect on water claims or rights in a particular location. Three statutes—P.L. 95-237, P.L. 96-560, and P.L. 103-77—have directed that the claims or rights to water and water projects on the Hunter and Fryingpan Rivers and their tributaries are to be unaffected by the wilderness designations in the laws. Two statutes designating wilderness areas along the lower Colorado River—P.L. 101-628 (AZ) and P.L. 103- 433 (CA)—specified that no right to Colorado River water was "expressly or impliedly" reserved. Two other statutes—P.L. 98-425 (CA) and P.L. 98-550 (WY)—specified no effect on water rights in one particular river and one specific river basin, respectively.

General Provisions

The explicit denial of reserved water rights associated with the wilderness designations has been included in 10 wilderness statutes. In four statutes—P.L. 103-433 (CA), P.L. 106-76 (CO), P.L. 111-11, Subtitle F (ID), and P.L. 111-11, Subtitle N (CO)—the denial of the reserved right is the extent of the provision. In the other six, the statutes also direct no effect on water agreements and/or state authorities. One additional statute—P.L. 103-77 (CO)—does not deny a reserved water right, but does prohibit federal assertion of and administrative and judicial consideration of any water claims.

Wilderness Study and Release

Intended or Potential Wilderness

Beginning with P.L. 94-357, the Alpine Lakes Area Management Act of 1976 (WA), Congress has enacted 17 wilderness statutes with intended or potential wilderness. These are areas within or adjoining designated wilderness areas that are to become wilderness when certain conditions have been met. In at least five statutes, areas are to be added to the designated wilderness when the specified nonfederal lands have been acquired. In all cases but one, the areas are to become wilderness when current prohibited or inconsistent uses have ceased and/or when incompatible conditions have been remediated. In all but two of the statutes, a *Federal Register* notice that the statutory conditions have been met is required before the area is officially added to the designated wilderness.

Wilderness Study and Review

A substantial number of wilderness statutes have directed the agencies to review the wilderness potential of certain lands and to present recommendations of wilderness designations to the President and to Congress. The Wilderness Act directed the Secretary of Agriculture to review the administratively identified national forest primitive areas within 10 years, with a third of the reviews completed within three years and the second third completed within seven years.[9] The act also directed the Secretary of the Interior to review all roadless areas of 5,000 acres or more within National Park System and National Wildlife Refuge System lands; recommendations were to be completed within 10 years, with a third done within three years and another third within seven years. A similar direction to review the wilderness potential of BLM lands was enacted in § 603 of the Federal Land Policy and Management Act of 1976 (FLPMA);[10] BLM wilderness recommendations were to be presented to the President within 15 years (i.e., by 1991) and to Congress not more than two years later. Questions have been raised about the legitimacy of BLM wilderness reviews of areas not originally identified as wilderness study areas (WSAs) under FLPMA.[11]

A total of 27 additional statutes directed the review of the wilderness potential of identified lands. About two-thirds of the statutes specified a deadline for presenting recommendations, commonly two, three, or five years. Two of the laws were only wilderness review statutes, and did not designate any wilderness areas. Two additional statutes repealed previously enacted wilderness study provisions, after the studies were completed, thus effectively providing release from the interim management guidelines (discussed below). The statutorily required wilderness reviews have all been completed, and recommendations have been presented to Congress; some agency wilderness recommendations remain pending.

Wilderness reviews of national forest lands have been and continue to be controversial. The Multiple-Use Sustained-Yield Act of 1960[12] explicitly identifies "wilderness" as an acceptable use for national forest lands. The National Forest Management Act of 1976 (NFMA)[13] requires periodically revised land management plans for the national forests that:

> provide for multiple use and sustained yield of the products and services obtained therefrom in accordance with the Multiple-Use Sustained-Yield Act of 1960, and, in particular, include coordination or outdoor recreation, range, timber, watershed, wildlife and fish, and wilderness. (16 U.S.C. § 1604(e)(1))

The periodic review of potential national forest wilderness in NFMA planning was modified in 1977 to accelerate the wilderness review portion of the planning process. In January 1979, the USFS issued nationwide recommendations on more than 60 million acres of land—some areas were recommended for wilderness, some for non-wilderness uses, and some to be examined further in the ongoing planning. The Roadless Area Review and Evaluation (RARE II)[14] was successfully challenged by the State of California on procedural grounds and vacated, raising questions about the management of lands that had been recommended for non-wilderness uses.[15]

Management During and after a Wilderness Review

The Wilderness Act and most of the initial statutory wilderness review provisions were silent on the management of the areas during and after the review. The Eastern Wilderness Act, P.L. 93- 622, and two other statutes (P.L. 94-577 and P.L. 105-277) directed that the wilderness characteristics of the areas under review were to be protected "until Congress determined otherwise," but only for a specified period after the recommendations were submitted (one through the third succeeding Congress, one for four years, and one until December 21, 2003). P.L. 94-199 simply directed that the wilderness character of the areas be protected. P.L. 96-550 was the first wilderness statute to require protection until Congress determined otherwise, without limitation, following the language in § 603(c) of FLPMA for the BLM wilderness study areas (WSAs). This language was used in seven other wilderness statutes. One law, P.L. 96-560, was particularly complicated—it provided the "until Congress determines otherwise" language for 10 areas, but directed grazing and mineral activities under laws generally applicable to national forests; it also limited the "until Congress determines otherwise" language to two years for one area, but directed that the Wilderness Act provisions on minerals apply to that area.

Because of the successful litigation over RARE II, many were concerned that, for areas recommended for non-wilderness uses, planned activities might be prevented if they were inconsistent with the Wilderness Act management guidelines (discussed below). A legislative provision, called *release language*, was developed to address this concern.[16] In general, release language provided that RARE II was sufficient for congressional deliberations over wilderness designation and that, in developing the first NFMA plan for a national forest, the USFS was not required to protect the wilderness characteristics of areas not designated. RARE II wilderness bills with release language were generally developed to address all the national forest lands (and occasionally some other federal lands) in a particular state. In total, 30 statewide national forest wilderness bills with release language were enacted between 1980 and 1990.[17] Because the initial NFMA plans have all been completed, release language is no longer relevant for national forest wilderness legislation.

Because FLPMA required the BLM to preserve the wilderness characteristics of its WSAs until Congress determines otherwise, release language for the WSAs not designated as wilderness is particularly important. It is also more controversial, because the BLM land and resource management planning process established in FLPMA does not mention wilderness and does not require periodic revision of the plans. Thus, release for BLM WSAs could be virtually permanent, in contrast to the periodic review required for national forest roadless areas. Since the BLM wilderness recommendations were made for each western state (because the BLM is organized into state offices), it is perhaps surprising that only one statewide BLM wilderness statute with release language (in contrast to statewide 30 USFS statutes with release language) has been enacted—P.L. 101-628, the Arizona Desert Wilderness Act of 1990. Release language for specific BLM WSAs has been enacted in 15 wilderness statutes.

PROHIBITED USES

The Wilderness Act, directly and by cross-reference in virtually all subsequent wilderness statutes, generally prohibits commercial activities, motorized uses, and roads, structures, and facilities in units of the National Wilderness Preservation System designated by acts of Congress. Specifically, § 4(c) states:

> Except as specifically provided for in this Act, and subject to existing private rights, there shall be no commercial enterprise and no permanent road within any wilderness area designated by this Act and, except as necessary to meet minimum requirements for the administration of the area for the purpose of this Act (including measures required in emergencies involving the health and safety of persons within the area), there shall be no temporary road, no use of motor vehicles, motorized equipment or motorboats, no landing of aircraft, no other form of mechanical transport, and no structure or installation within any such area.

This section thus prohibits most businesses, except "for activities which are proper for realizing the recreational or other wilderness purposes of the areas" (§ 4(d)(6)). This section effectively prohibits development of commercial resources, such as timber, although the Wilderness Act permits livestock grazing and some mineral development, as discussed below.

The Wilderness Act also prohibits motorized or mechanical entry, via cars, trucks, off-road or all-terrain vehicles, bicycles, aircraft, or motorboats, except in emergencies and in specified circumstances, as discussed below. Finally, the act prohibits human infrastructure—roads, buildings, dams and pipelines, and such—from wilderness areas, although the act is silent on the treatment (e.g., maintenance or removal) of infrastructure within designated wilderness areas; many of the provisions in wilderness statutes addressing nonconforming uses address existing and potential infrastructure in designated areas.

Area Closures to Public Access

In addition, several wilderness statutes authorized closing certain wilderness areas to some public access. Five statutes authorized temporarily closing areas to hunting, fishing, and trapping for public safety, administration, and use and enjoyment of the areas.[18] Two statutes—P.L. 109-362 and P.L. 111-11, Subtitle L—authorized temporary closures for tribal religious or cultural needs and activities. P.L. 106-65 authorized closure of a portion of a wilderness abutting the Mexican border for public safety and national security. Finally, P.L. 107-107 directed that the designated wilderness area be closed to the public, because it had been the "Live Impact Area" of the Vieques Naval Training Range and could still contain some live munitions.

Withdrawal from Public Land, Mining, and Mineral Leasing Laws

Withdrawal in this context is the term used for preventing an individual from obtaining title to federal lands or resources under the various laws that have historically permitted such actions.

Public Land Laws

Public land laws refer to the array of statutes that allowed individuals to obtain title to unreserved public domain lands—that is, to federally owned lands that were obtained by purchase or treaty from a sovereign power (the public domain) and that had not been reserved for a particular purpose by Congress. The best known of the public land laws was the Homestead Act, which allowed settlers to obtain title to 160 acres of unreserved public domain lands by agreeing to develop the lands and live on them for at least five years.

The Wilderness Act is silent on the application or withdrawal of areas designated from the public land laws. This is likely because the areas designated were all within national forests, which were already withdrawn from the public land laws as congressionally authorized reservations of land. Similarly, therefore, all wilderness designations are automatically withdrawn from the public land laws, because the designations are implicitly congressional reservations of federal land. Furthermore, the vast majority of the public land laws were repealed in Title VII of the Federal Land Policy and Management Act of 1976 (FLPMA, P.L. 94-579). Nonetheless, 14 wilderness statutes, beginning with P.L. 106-353 in 2000, still have explicitly withdrawn the designated areas from disposal under the public land laws.

Mining and Mineral Leasing Laws

The General Mining Law of 1872 (30 U.S.C. Chapter 2) established a system of relatively free access for individuals and corporations to prospect for hardrock (or locatable) minerals (e.g., gold, silver, copper) on open federal lands and to stake a claim on the deposit. The minerals can then be extracted from sites with valid mining claims. The claim can be "patented" to transfer title to the land to the claimant, although patenting the land is not required to extract the minerals.

Mineral leasing for "coal, phosphate, sodium, potassium [potash], oil, oil shale, gilsonite (including all vein-type solid hydrocarbons), or gas" under federal lands is governed by the Mineral Leasing Act of 1920 (30 U.S.C. §§ 181 et seq.) and the Mineral Leasing Act for Acquired Lands (31 U.S.C. §§ 351 et seq.). Leases are generally offered to the public competitively, and establish a right to extract the leased mineral in exchange for payment of fees and royalties. In addition, the Geothermal Steam Act of 1970 (P.L. 91-581; 30 U.S.C. §§ 1001 et seq.) allows geothermal energy production from federal lands nominated for leasing. The Geothermal Steam Act does not exclude wilderness areas from lands that can be nominated for leasing.

The Wilderness Act allowed continued prospecting in wilderness areas (§ 4(d)(2)). It also allowed filing claims to hardrock minerals and establishing contracts for leasable minerals within wilderness areas "until midnight on December 31, 1983." As discussed below, the Wilderness Act included guidelines for administering valid mineral rights. Since 1978, 18 wilderness statutes have explicitly withdrawn the designated areas from availability under the General Mining Law. These statutes also withdrew the designated areas from availability under the Mineral Leasing Act and Geothermal Steam Act, except for P.L. 95-495 (withdrawn under neither law), P.L. 96-428 (not withdrawn from geothermal leasing), and P.L. 111-11, Subtitle G (not withdrawn from mineral leasing). These provisions seem to be becoming more common, as 6 of the 16 wilderness subtitles of P.L. 111-11 and five of the six laws enacted in the 109th Congress withdraw the lands.

Mineral Materials

The Materials Act of 1947 (30 U.S.C. §§ 601 et seq.) authorizes the sale of common variety mineral materials—sand, gravel, crushed rock, and others—from federal lands. Beginning in 1996, 12 wilderness statutes have withdrawn the designated lands from disposal of common variety mineral materials.

NONCONFORMING PERMITTED USES

The Wilderness Act and a large number of subsequent wilderness statutes allow uses, activities, or infrastructure that do not conform with the general prohibitions on commercial activities, motorized access, and infrastructure. Many of these nonconforming permitted uses were explicitly allowed in the Wilderness Act, including access for management and emergencies and access for inholdings, as well as activities for controlling fires and insect and disease infestations, continued motorized access, livestock grazing, and water project developments. Subsequent statutes have expanded on these provisions and have addressed additional concerns, such as fish and wildlife management activities, development or maintenance of and access to certain existing and potential infrastructure, military overflights of wilderness areas, mineral activities, and access for other specific purposes.

Access for Management Requirements and Emergencies

The Wilderness Act, in the provision cited above, explicitly allowed motorized access for "minimum management requirements" and "emergencies involving health and safety."

Interestingly, neither Congress nor the agencies have defined the *minimum management requirements* for which motorized access is allowed. One statute, P.L. 90-352, reiterated the

Wilderness Act provision. In addition, P.L. 98-425 explicitly allowed motorized access on a particular fire road between two designated wilderness areas for administrative purposes. Two statutes—P.L. 98-406 and P.L. 101-628—modify authorized uses of particular roads within the designated wilderness. No other wilderness statutes have addressed access for management or emergencies.

Access to Nonfederal Inholdings

Inholdings are nonfederally owned lands—lands owned by state or local governments or by a private individual or entity—that are contained within (surrounded by) a federal unit, such as a national park, national forest, national wildlife refuge, or designated wilderness. The Wilderness Act explicitly directs that nonfederal landowners "be given such rights as may be necessary to assure adequate access" to their lands. The act also provides reasonable access for valid mining claims and other valid occupancies, requiring that the Secretary "shall, by reasonable regulations consistent with the preservation of the area of wilderness, permit ingress and egress to such surrounded areas by means which have been or are being customarily enjoyed with respect to other such areas similarly situated."

Subsequent wilderness statutes have generally followed the Wilderness Act provisions. The first access provision, in P.L. 98-406, allowed limited access for private and administrative purposes along one route into one designated wilderness areas. Ten additional statutes have paralleled the Wilderness Act, directing reasonable access or adequate access for reasonable use and enjoyment of the surrounded nonfederal lands. One statute, P.L. 106-353, also included the right of public access, including for commercial vehicles, to one area.

Fire, Insect, and Disease Control

The Wilderness Act specifically authorized exceptions to the general prohibitions on activities in wilderness areas for controlling fires and insect and disease infestations; § 4(d)(1) allowed that "such measures may be taken as may be necessary in the control of fire, insects, and diseases, subject to such conditions as the Secretary deems desirable." This language has been cited or replicated in 23 subsequent wilderness statutes, beginning in 1978. It has become more common in recent years; for example, such a provision was included in 8 of the 16 wilderness subtitles of P.L. 111-11. In three wilderness statutes, control activities are described as appropriate for watershed protection, and in one for the necessary control of southern pine beetles. Finally, one statute—P.L. 96-560—required a review of the practices, policies, and regulations to control diseases, insects, and forest fires to insure timely and efficient control; reasonable protection of adjacent lands; and conformance with the intent of Congress.

Continued Motorized Uses

Despite prohibiting motorized access to designated wilderness areas, the Wilderness Act also explicitly directed, in § 4(d)(1), that "the use of aircraft or motorboats, where these uses have already become established, may be permitted to continue subject to such restrictions as the Secretary ... deems desirable." Several subsequent wilderness statutes have provided explicit guidance on continued motorized access. Most commonly (five statutes), the provisions allow motorboats in particular lakes or streams; in at least one instance, the size of the motors that can be used is restricted. Two statutes—P.L. 95-495 and P.L. 96-487—authorized motorized portages between lakes. Two statutes—P.L. 95-495 and P.L. 110-229—specifically authorized continued aircraft use within the areas. (Military aircraft flights over wilderness areas are discussed separately, below.) Finally, two statutes—P.L. 95-495 and P.L. 108-447—authorized the continued use of snowmobiles in specific areas. As can be seen from this brief summary, P.L. 95- 495, the Boundary Waters Canoe Area Wilderness, contained more of these types of provisions than other wilderness statutes; P.L. 95-495 is also the only wilderness statute that amended the Wilderness Act, adding a new § 4(d)(5) addressing management of the areas, including continued motorboat use.

Livestock Grazing

Continued livestock grazing in wilderness has attracted substantial attention. The Wilderness Act, in § 4(d)(4)(2), specifically directed that "the grazing of livestock, where established prior to the effective date of this Act, shall be permitted to continue subject to such reasonable regulations as are deemed necessary by the Secretary of Agriculture."

Despite this language, many have expressed concerns that wilderness designations will reduce historic livestock grazing. The concerns were sufficient for the House Committee on Interior and Insular Affairs to include a discussion of the congressional intent for continued livestock grazing management in wilderness areas in the committee report on P.L. 96-560—H.Rept. 96-617. Continued concerns led the House Committee on Interior and Insular Affairs to expound still further on the congressional intent for continuing livestock grazing in designated wilderness areas. The committee included "Appendix A—Grazing Guidelines" in H.Rept. 101-405 that accompanied the bill that was enacted as P.L. 101-628. (Summaries of the relevant portions of the committee reports are included following the respective statutes in Appendix B.)[19]

A total of 33 subsequent wilderness statutes have expressly directed continued livestock grazing in conformance with the Wilderness Act. The majority (27 statutes) have also directed that grazing be managed in conformance with § 108 of P.L. 96-560; with the relevant portion of H.Rept. 96-617; with § 101(f) of P.L. 101-628; and/or with Appendix A of H.Rept 101-405. Several (at least 11 statutes) have modified the continuation to be "subject to such reasonable regulations, policies, and practices as are deemed necessary." Such provisions seem to be becoming more common; 9 of the 16 wilderness subtitles in P.L. 111-11 and four of the six wilderness statutes enacted in the 109th Congress included grazing management provisions.

A few wilderness statutes have contained additional provisions. Five statutes directed the agencies to conduct a study to insure that the grazing regulations, policies, and practices were in accordance (or in conformance) with the Wilderness Act. A sixth statute (P.L. 111-11, Subtitle F) directed an inventory of grazing-related facilities and improvements, and acceptance of donations with subsequent cancellation of grazing permits within the wilderness areas.[20] Two statutes (P.L. 96-550 and P.L. 111-11, Subtitle F) explicitly allowed additional fencing within the wilderness, while another (P.L. 98-425) explicitly allowed motorized access for livestock management.

Access for Water Infrastructure

Numerous wilderness statutes have provided for the maintenance and reconstruction of existing water resource infrastructure (dams, canals, pipelines, etc.) and construction of new infrastructure in some instances, usually with permission for the necessary motorized access to accomplish these actions. The Wilderness Act allowed the President to permit such activities:

> The President may, within a specific area and in accordance with such regulations as he may deem desirable, authorize prospecting for water resources, the establishment and maintenance of reservoirs, water-conservation works, power projects, transmission lines, and other facilities needed in the public interest, including the road construction and maintenance essential to development and use thereof, upon his determination that such use or uses in the specific area will better serve the interests of the United States and the people thereof than will its denial. (§ 4(d)(4)(1)

A total of 13 subsequent wilderness statutes have provided for the construction, operation, and maintenance of water resource facilities, for one specific area in some statutes and for all the areas designated in other statutes. In contrast to many other provisions, the language in each of these statutes appears to be unique, without clear repetition of the same language. In at least two statutes—P.L. 95-237 and P.L. 98-428—helicopter access is allowed for sanitary facilities within wilderness areas.

In addition, six wilderness statutes effectively prohibit new water resource development within the wilderness areas. The language is virtually identical in the six statutes: "neither the President nor any other officer, employee, or agent of the United States shall fund, assist, authorize, or issue a license or permit for the development of any new water resource facility within the wilderness areas designated by this Act."

Access for Other Infrastructure

Eleven wilderness statutes allow for the construction and maintenance of, and generally limited motorized access in support of, facilities for hydrological, meteorological, climatological, atmospheric, and/or telecommunications needs. The language used is often similar, but is also usually distinct in some manner. In some instances, the infrastructure is allowed if it is essential to flood control and reservoir operations or for research; in other cases, it must be appropriate for science, education, or conservation purposes. Some statutes direct reasonable or prescribed terms and conditions for the infrastructure and/or access.

Another nine wilderness statutes provide for the maintenance of specific facilities, and access for such activities. Two statutes direct the maintenance of specific trails and associated structures; another allows the maintenance of cabins in specific areas. One directs the conversion of roads to trails. One statute allows operation and maintenance of a lookout tower; another allows motorized access for the removal of a fire tower. Two statutes allow the operation, maintenance, upgrade, and replacement of power lines through specific areas. Finally, one statute authorizes the right-of-way and construction of a space laser energy facility within 15 years (now expired, without such construction).

Access for Fish and Wildlife Management Activities

As noted above, the Wilderness Act and several subsequent wilderness statutes direct that the wilderness designations are to have no effect on state jurisdiction or responsibilities for fish and wildlife. As also noted above, three wilderness statutes authorize periods when or zones where the wilderness may be closed to hunting, fishing, and trapping, for safety and administrative reasons.

Several statutes have also provided for facilities or motorized access in support of state fish and wildlife management. In six instances, the provisions apply only to one area or activity. The House Committee on Interior and Insular Affairs report, H.Rept. 101-405, accompanying the bill enacted as P.L. 101-628, included a lengthy Appendix B—Wildlife Management Guidelines. This committee report provides detailed guidance on the committee's statement of congressional intent regarding motorizes access, facilities, and state agency activities for fish and wildlife management within wilderness areas; the statute itself contains no direction on access for fish and wildlife management activities. (A summary of this appendix follows P.L. 101-628 in Appendix B of this report.)

In contrast to the Grazing Guidelines (Appendix A) of H.Rept. 101-405, the Wildlife Management Guidelines have not been referenced in subsequent wilderness statutes. Instead, five statutes—P.L. 107-282 (NV); P.L. 108-424

(NV); P.L. 109-432 (NV); P.L. 111-11, Subtitle F (ID); and P.L. 111-11, Subtitle O (UT)—contain lengthy provisions directing state fish and wildlife management within the designated wilderness areas. These provisions are similar to, but less detailed than, H.Rept. 101-405, Appendix B—Wildlife Management Guidelines.

Low-Level Military Overflights

A concern about the limitations of wilderness designations on military air training arose in the late 1980s. P.L. 101-195, the Nevada Wilderness Protection Act of 1989, included a provision directing that the statute would have no effect on low-level military overflights, new units of special use airspace, or military flight training routes over four specific wilderness areas. Similar language has been included in 11 subsequent wilderness statutes, primarily in Nevada and southern California, but also in Arizona, Idaho, and Utah.

Access for Minerals Activities

As discussed above, the Wilderness Act (and many subsequent wilderness statutes) withdrew designated wilderness areas from access under the mining and mineral leasing laws after December 31, 1983. However, § 4(d)(2) of the act allows mineral prospecting, as long as it is conducted "in a manner compatible with the preservation of the wilderness environment." In addition, for the nearly 20 years after its enactment, § 4(d)(3) of the Wilderness Act allowed establishing and developing valid mineral rights in wilderness areas,

> subject, however, to such reasonable regulations governing ingress and egress as may be prescribed by the Secretary of Agriculture consistent with the use of the land for mineral location and development and exploration, drilling, and production, and use of land for transmission lines, waterlines, telephone lines, or facilities ... including where essential the use of mechanized ground or air equipment.... Mineral leases, permits, and licenses ... shall contain such reasonable stipulations as may be prescribed by the Secretary of Agriculture for the protection of the wilderness character of the land consistent with the use of the land for the purposes for which they are leased, permitted, or licensed.

Several subsequent wilderness statutes have also addressed minerals. Two—P.L. 95-237 and P.L. 98-428—directed continued development of mineral information within the designated areas. Another two statutes—P.L. 96-312 and P.L. 98-425—provided for cobalt extraction from within specific wilderness areas. Finally, P.L. 106-456 allowed the historic uses of an existing mine road within a wilderness area.

Access for Other Specific Activities

As discussed above under "Prohibited Uses," the Wilderness Act generally prohibited commercial activities within wilderness areas, except in support of recreational and other appropriate uses of wilderness areas. This direction was restated in the reverse—allowing commercial activities that support recreational and other wilderness purposes—in P.L. 96-312 and P.L. 111-11, Subtitle F. In addition, Subtitle K and Subtitle M of P.L. 111-11 addressed concerns about the implementation of this provision. Subtitle K directed that outfitter and guide permits for the additional designated wilderness areas were to be added to the existing limits for wilderness. Both subtitles directed that the wilderness designation was not to reduce riding or pack stock use or impose additional use constraints in the areas, in accordance with the Wilderness Act and "subject to any terms and conditions determined to be necessary by the Secretary." Finally, P.L. 111-11, Subtitle A, authorized a competitive event through the designated wilderness to continue.

Various wilderness statutes have authorized access to specific areas for other specific purposes. Two statutes—P.L. 97-384 and P.L. 101-633—authorized access to cemeteries that were within the designated wilderness areas. Three wilderness statutes addressed rights-of-way; one (P.L. 95- 249) directed no effect on two rights-of-way claims, another (P.L. 98-425) authorized a powerline right-of-way if an application were made within 10 years (it was not), and the third (P.L. 103-433) authorized a right-of-way for the Department of Defense in one area. Finally, five statutes addressed access for Native American activities. Three—P.L. 107-282, P.L. 108-424, and P.L. 109-432—directed no diminution of tribal access to wilderness areas. Another (P.L. 109-362) directed access in accordance with the Wilderness Act (which was silent on the issue of tribal access) and the American Indian Religious Freedom Act.[21] The fifth statute (P.L. 111-11, Subtitle L) allowed for Native American access to a specific areas. P.L. 109-362 and P.L. 111-11, Subtitle L also allowed temporary closures for tribal needs, as noted above.

APPENDIX A. LIST OF WILDERNESS STATUTES CONTAINING THE VARIOUS STATUTORY PROVISIONS AND GUIDELINES ON PROHIBITED AND PERMITTED USES

This appendix identifies the specific section in each of the wilderness statutes containing the provisions specified. Because of the 16 subtitles in P.L. 111-11, the subtitle letter is shown following the statute number to assist in finding the provision (with IIE used for Subtitle E of Title II; the other subtitles are all in Title I). The provisions are summarized or quoted for each statute, in chronological order, in Appendix B.

Statutory Provisions

General Management Guidance

Manage in Accordance with the Wilderness Act (P.L. 88-577)—All Statutes Except:			
P.L. 90-532	P.L. 90-544	P.L. 92-476	
	Prepare a Management Plan		
P.L. 95-237, § 2(c)	P.L. 96-312, § 5(a)	P.L. 106-353, § 6(h)	P.L. 111-11, F, § 1503(b)(5)

No Buffer Zones

P.L. 96-550, § 105	P.L. 96-560, § 110	P.L. 98-328, § 6	P.L. 98-339, § 9
P.L. 98-406, § 101(d)	P.L. 98-428, § 303	P.L. 98-508, § 7	P.L. 98-550, § 504
P.L. 98-585, § 9	P.L. 98-586, § 7	P.L. 100-184, § 7	P.L. 100-499, § 7
P.L. 100-668, Title IV (a)(3)	P.L. 101-195, § 7	P.L. 101-401, § 6	P.L. 101-628, § 101(d)
P.L. 101-633, § 6	P.L. 102-301, § 3(d)	P.L. 103-77, § 3(e)	P.L. 103-433, § 103(d)
P.L. 106-145, § 8	P.L. 106-353, § 6(i)	P.L. 107-282, § 204	P.L. 108-424, § 205
P.L. 109-163, § 384(h)	P.L. 109-362, § 4(l)	P.L. 109-432, § 325	P.L. 111-11, C, § 1202(f)
P.L. 111-11, F, § 1503(b)(10)	P.L. 111-11, L, § 1851(b)(3)	P.L. 111-11, O, § 1972(b)(4)	

No Effect on State Fish and Wildlife Jurisdiction and Responsibilities

P.L. 88-577, § 4(d)(8)	P.L. 95-495, § 14	P.L. 96-312, § 7(c)	P.L. 98-322, § 104(b)
P.L. 98-406, § 101(e)(2)	P.L. 98-550, § 503	P.L. 98-585, § 5(b)	P.L. 100-184, § 8
P.L. 100-499, § 18	P.L. 101-195, § 9	P.L. 101-401, § 8	P.L. 101-628, § 101(e)
P.L. 101-633, § 7	P.L. 103-77, § 3(c)	P.L. 103-433, § 103(e)	P.L. 106-76, § 8(c)
P.L. 107-282, § 208(a)	P.L. 107-370, § 3(c)	P.L. 108-424, § 209(a)	P.L. 109-94, § 3(h)
P.L. 109-163, § 384(f)	P.L. 109-362, § 4(j)(2)	P.L. 109-382, § 104(c) & § 213(b)	P.L. 109-432, § 329(a)
P.L. 111-11, A, § 1001(e) P.L. 111-11, G, § 1602(c)(4)	P.L. 111-11, C, § 1202(g) P.L. 111-11, K, § 1803(i)(2)	P.L. 111-11, E, § 1405(c)(4) P.L. 111-11, O, § 1972(b)(10)	P.L. 111-11, F, § 1503(b)(8)(A) P.L. 111-11, IIE, § 2405(i)

No Effect on Other Jurisdiction and Authorities

P.L. 95-495, §§ 16 & 17 P.L. 106-65, § 3032 P.L. 111-11, K, § 1802(1)	P.L. 98-550, § 505 P.L. 106-145, § 6(b) P.L. 111-11, L, § 1851(b)(3)	P.L. 101-628, § 301(g) P.L. 107-282, § 211	P.L. 103-433, § 103(g) P.L. 111-11, H, § 1653(b)

Land & Rights Acquisition

P.L. 88-577, § 5(c) P.L. 98-574, § 5 P.L. 104-333, § 1022(b)(4)	P.L. 93-622, § 6(b) P.L. 100-184, § 9	P.L. 97-466, § 4 P.L. 101-628, § 101(i)	P.L. 98-425, § 103(c) P.L. 103-77, § 2(a)(8)

Water Rights

	No Effect on State Authorities and Water Agreements		
P.L. 95-495, § 15 P.L. 103-433, § 203 P.L. 108-447, § 140(d)(8) P.L. 111-11, IIE, § 2405(h)	P.L. 96-550, § 102(a)(9) P.L. 106-353, § 6(l)(2) P.L. 109-94, § 3(i)	P.L. 101-628, § 302 P.L. 107-282, § 203(d) P.L. 109-432, § 324(d)	P.L. 103-77, § 8(b) P.L. 108-424, § 204(d) P.L. 111-11, O, § 1972(b)(9)

	Neither Claim Nor Denial of Claim		
P.L. 88-577, § 4(d)(7) P.L. 98-550, § 502	P.L. 96-312, § 7(b) P.L. 106-399, § 203	P.L. 98-406, § 101(e)(1)	P.L. 98-428, § 302(a)
	Reserved Water Rights		
P.L. 100-668, § 502 P.L. 103-433, § 706	P.L. 101-195, § 8 P.L. 107-370, § 3(d)	P.L. 101-628, §§ 101(g) & 301(c)	P.L. 102-301, § 3(e)
	No Claim on Water—Geographic-Specific Provisions		
P.L. 95-237, § 2(e) P.L. 101-628, § 301(c)	P.L. 96-560, § 102(a)(5) P.L. 103-77, §§ 2(a)(13) & 8(g)(1)	P.L. 98-425, § 101(a)(25) P.L. 103-433, § 204	P.L. 98-550, § 201(c)
No Claim on Water—General Provisions			
P.L. 103-77, § 8(b)	P.L. 106-76, § 10	P.L. 106-353, § 6(l)(2)(A)	P.L. 107-282, § 203(d)
P.L. 108-424, §	P.L. 109-94, § 3(i)(A)	P.L. 109-432, §	P.L. 110-11F, § 1503(b)(12)
204(d)(2)(A)		324(d)(3)(A)	
P.L. 111-11, N, § 1952(f)	P.L. 111-11, O, § 1972(b)(9)		

Wilderness Study and Release

Intended or Potential Wilderness			
P.L. 94-357, §§ 3(d) & (e)	P.L. 94-544, § 1	P.L. 94-567, § 3	P.L. 95-625, § 403
P.L. 97-250, § 2(a)	P.L. 98-425, §§ 106 & 108	P.L. 100-524, § 2(b)	P.L. 100-668,Title IV (a)(2)
P.L. 102-301, § 2(5)	P.L.103-433, § 601(b)	P.L. 109-362, § 6	P.L. 111-11, B, § 1103
P.L. 111-11, C, § 1202(c)	P.L. 111-11, I, § 1702(c)	P.L. 111-11, L, § 1853(c)	P.L. 111-11, M, § 1902
P.L. 111-11, N, § 1952(c)			
	Wilderness Study and Review		
P.L. 88-577, §§ 3(b) & (c)	P.L. 90-544, § 604	P.L. 92-400, § 5	P.L. 92-476, § 5(b)
P.L. 92-521, § 4	P.L. 92-528, § 1(a)	P.L. 93-622, § 4 & § 6(a)	P.L. 94-199, § 8

(Continued).

P.L. 94-557, §§ 3(b) & (d)	P.L. 94-567, § 8	P.L. 94-579, § 603	P.L. 95-150
P.L. 96-487, § 704	P.L. 96-550, § 103	P.L. 96-560, §§ 105 & 106	P.L. 98-323, § 104(e)
P.L. 98-324, § 6	P.L. 98-406, § 102(a)	P.L. 98-425, §§ 102 & 111(e)	P.L. 98-430, § 2
P.L. 98-514, § 5(e)	P.L. 98-550, § 301	P.L. 98-578, § 6	P.L. 98-586, § 6
P.L. 101-628, § 102	P.L. 103-433, § 105	P.L. 105-277, § 610	P.L. 106-399, § 204
P.L. 111-11, B, § 1102(B)			
Management During and After Review and Release in a Statewide Statute			
P.L. 96-487, § 708	P.L. 96-550, § 104	P.L. 96-560, § 107	P.L. 97-384, § 4
P.L. 97-407, § 5	P.L. 97-466, § 5	P.L. 98-321, § 5	P.L. 98-322, § 105
P.L. 98-323, § 104	P.L. 98-324, § 5	P.L. 98-328, § 7	P.L. 98-339, § 5
P.L. 98-406, § 103	P.L. 98-425, § 111	P.L. 98-428, § 201	P.L. 98-430, § 6
P.L. 98-508, § 4	P.L. 98-514, § 5	P.L. 98-515, § 5	P.L. 98-550, § 401
P.L. 98-574, § 5	P.L. 98-585, § 8	P.L. 98-586, § 5	P.L. 99-197, § 4
P.L. 99-504, § 104	P.L. 100-184, § 6	P.L. 100-499, § 6	P.L. 100-547, § 202
P.L. 101-195, § 5	P.L. 101-401, § 5	P.L. 101-628, § 102	
Management During and After Review and Release of Specific Areas			
P.L. 98-140, §§ 3(a)-(c) P.L. 102-301, § 5	P.L. 98-578, § 5 P.L 103-77, § 4	P.L. 99-490, § 5 P.L.103-433, § 104(a)	P.L. 100-326, § 4 P.L. 106-145, § 5
P.L. 107-282, § 207 P.L. 109-362, § 5	P.L. 108-424, § 208 P.L. 109-432, § 328	P.L. 109-94, § 3(f) P.L. 111-11, F, § 1503(a)(1)	P.L. 109-163, § 384(i) P.L. 111-11, G, § 1602(e)
P.L. 111-11, I, § 1703	P.L. 111-11, J, § 1753	P.L. 111-11, K, § 1804	P.L. 111-11, O, § 1972(c)

Prohibited Uses

General Prohibitions

P.L. 88-577, § 4(c)	P.L. 90-532, § 3		

Area Closures

P.L. 93-429, § 3 P.L. 107-282, § 208(e)	P.L. 106-65, § 3032(i) P.L. 108-424, § 209(e)	P.L. 106-353, § 6(f) P.L. 109-362, § 4(k)	P.L. 107-107, § 1049 P.L. 109-432, § 329(e)
P.L. 111-11, L, § 1853(d)(6)			

Withdrawal from Public Land, Mining, Mineral Leasing, Geothermal Leasing, and/or Mineral Materials Laws

P.L. 88-577, § 4(d)(3)	P.L. 95-495, § 14	P.L. 96-428, § 1	P.L. 98-425, § 306(a)(1)
P.L. 104-333, § 1022(b)(4)	P.L. 106-353, § 6(c)	P.L. 107-282, § 202(d)	P.L. 108-424, § 203(d)
P.L. 109-94, § 3(j) P.L.109-432, § 323(d)	P.L. 109-163, § 384(b) P.L. 111-11, C, § 1202(i)	P.L. 109-362, § 4(d) P.L. 111-11, F, § 1503(b)(2)	P.L. 109-382, § 104(d) P.L. 111-11, G, § 1602(d)
P.L. 111-11, K, § 1803(d)	P.L. 111-11, L, § 1853(d)(3)	P.L. 111-11, IIE, § 2405(a)	

Notes:

P.L. 88-577, P.L. 95-495, P.L. 96-428, P.L. 98-425, and P.L. 104-333 did not explicitly withdraw lands under the public land laws.

P.L. 95-495 and P.L. 111-11, Subtitle G did not explicitly withdraw lands from the mineral leasing laws.

P.L. 88-577, P.L. 96-428, P.L. 98-425, P.L. 108-424, P.L. 109-382, and P.L. 109-432 did not explicitly withdraw lands from the mineral materials laws.

Nonconforming Permitted Uses

Access for Management Requirements and Emergencies

P.L. 88-577, § 4(c)	P.L. 90-352, § 3	P.L. 98-406, § 101(a)(23)	P.L. 98-425, § 101(a)(6)
P.L. 101-628, § 101(k)			

Access to Nonfederal Inholdings

P.L. 88-577, § 5(a) & (b)	P.L. 98-406, § 101(a)(23)	P.L. 100-184, § 9	P.L.106-353, § 9
P.L. 106-399, § 202(c)	P.L. 106-456, § 3(b)	P.L. 109-362, § 4(f)	P.L.110-229, § 101(c)(2)
P.L. 111-11, F, §	P.L. 111-11, G, § 1602(c)(5)	P.L. 111-11, I, § 1702(b)(4)	P.L. 111-11, K, § 1803(f)
1503(b)(7)			

Fire, Insect, and Disease Control

P.L. 88-577, § 4(d)(1)	P.L. 95-237, § 2(c)	P.L. 96-560, § 109	P.L. 98-425, § 103(b)(2)
P.L. 100-184, § 10	P.L. 100-499, § 16	P.L. 100-547, § 201(c)	P.L. 101-401, § 7
P.L. 101-633, § 8	P.L. 102-301, § 3(b)	P.L. 106-145, § 6(b)	P.L. 107-282, § 209
P.L. 107-370, § 4	P.L. 108-424, § 210	P.L. 109-362, § 4(e)	P.L. 109-432, § 330
P.L. 111-11, C, § 1202(h)	P.L. 111-11, E, § 1405(c)(2)	P.L. 111-11, F, § 1503(b)(9)	P.L. 111-11, K, § 1803(e)
P.L. 111-11, L, §	P.L. 111-11, N, § 1952(g)	P.L. 111-11, O, §	P.L. 111-11, IIE, §
1853(d)(4)		1972(b)(3)	2405(e)(3)

Continued Motorized Uses

P.L. 88-577, § 4(d)(1)	P.L. 93-429, § 2	P.L. 95-495, §§ 4 & 8	P.L. 96-312, § 7(a)(1)
P.L. 96-487, § 103(b)	P.L. 98-425, § 101(a)(7)	P.L. 98-430, § 1(4)	P.L. 102-301, § 2(5)
P.L. 108-447, § 140(d)(2)	P.L. 109-362, § 4(m)	P.L. 110-229, § 10(b)(3) & (4)	P.L. 111-11, H, §§ 1653(b) & 1654

Livestock Grazing

P.L. 88-577, § 4(d)(4)(2)	P.L. 96-312, § 7(a)(2)	P.L. 96-550, § 102(a)(5)	P.L. 96-560, § 108 & H.Rept. 96-617
P.L. 98-406, §§ 101(f)(1) & 302(b)	P.L. 98-425, §§ 101(a)(2) & (25) & 103(b)(1)	P.L. 98-428, § 301	P.L. 98-603, § 102(a)
P.L. 99-504, § 102(a)	P.L. 99-584, § 2(b)	P.L. 100-225, § 402(b)	P.L. 101-195, § 6(a)
P.L. 101-628, § 101(f) &	P.L. 103-77, § 3(b)	P.L. 103-433, § 103(c)	P.L. 106-353, § 6(g)(2)
H.Rept. 101-405, App. A			
P.L. 106-399, § 202(d)(1)	P.L. 106-554, § 8(d)	P.L. 107-282, § 203(b)	P.L. 107-370, § 3(b)
P.L. 108-424, § 204(b)	P.L. 109-94, § 3(g)	P.L. 109-163, § 384(g)	P.L. 109-362, § 4(b)
P.L. 109-432, § 324(b)	P.L. 111-11, E, § 1405(c)(3)	P.L. 111-11, F, § 1503(b)(3)	P.L. 111-11, G, § 1602(c)(3)
P.L. 111-11, I, § 1702(b)(3)	P.L. 111-11, J, § 1752(b)(3)	P.L. 111-11, K, § 1803(h)	P.L. 111-11, L, § 1853(d)(5)
P.L. 111-11, O, § 1972(b)(2)	P.L. 111-11, IIE, § 2405(b)(2)		

(Continued).

	Study of Grazing Regulations, Policies, and Practices		
P.L. 98-406, § 101(f)(2)	P.L. 98-428, § 301(b)	P.L. 98-550, § 501	P.L. 99-504, § 102(b)
P.L. 101-195, § 6(b)	P.L. 101-628, § 101(f)(2)		

Access for Water Infrastructure

P.L. 88-577, § 4(d)(4)(1)	P.L. 91-82, § 3	P.L. 95-237, § 2(i)	P.L. 95-495, § 13
P.L. 96-312, § 7(a)(4)	P.L. 96-560, § 102(a)(17)	P.L. 96-585, § 1(d)	P.L. 98-428, § 302(b)
P.L. 98-603, § 109	P.L. 100-668, § 202	P.L. 101-628, §§ 101(a)(3) & 101(l)	P.L. 103-77, §§ 8(d)(4), (e), & (f)
P.L. 107-370, § 7	P.L. 111-11, M, § 1902(1)(C)		
	Prohibition on New Water Infrastructure		
P.L. 103-77, § 8(c)	P.L. 107-282, § 203(d)(4)(B)	P.L. 108-424, § 204(d)(4)(B)	P.L. 109-94, § 3(i)(4)(B)
P.L. 109-432, § 324(d)(5)(B)	P.L. 111-11, IIE, § 2405(h)(3)		

Access for Other Infrastructure

Meteorological, Hydrological, and Related Infrastructure			
P.L. 98-406, § 101(a)(13)	P.L. 98-428, § 305	P.L. 101-195, § 10	P.L. 107-282, § 210
P.L. 108-424, § 211	P.L. 109-118, § 3(d)	P.L. 109-362, § 4(g)	P.L. 109-432, § 331
P.L. 110-229, § 101(b)(3)	P.L. 111-11, M, § 1903(c)	P.L. 111-11, O, § 1972(b)(8)	
	Other Infrastructure		
P.L. 98-322, § 104(c)	P.L. 100-547, § 201(d)	P.L. 100-668, §§ 102 & 302	P.L. 101-628, § 101(a)(20)

Meteorological, Hydrological, and Related Infrastructure			
P.L. 103-433, § 102(1)	P.L. 106-156, § 2(d)	P.L. 109-382, § 213(c)	P.L. 110-229, § 101(b)(5)
P.L. 111-11, M, § 1902(1)(B)			

Access for Fish and Wildlife Management Activities

P.L. 98-550, § 201(a)(11)	P.L. 101-628 H.Rept 101-405, App. B	P.L. 102-301, § 3(c)	P.L. 103-433, § 103(f)
P.L. 107-282, § 208	P.L. 108-424, § 209	P.L. 109-362, § 4(j)(1)	P.L. 109-382, § 213(b)
P.L. 109-432, § 329	P.L. 111-11, F, § 1503(b)(8)	P.L. 111-11, K, § 1803(i)(1)	P.L. 111-11, O, § 1972(b)

Low-Level Military Overflights

P.L. 101-195, § 11	P.L. 101-628, §§ 101(h) & 301(f)	P.L. 103-433, § 802	P.L. 107-282, § 205(a)
P.L. 107-370, § 5	P.L. 108-424, § 206	P.L. 109-362, § 4(h)	P.L. 109-432, § 326
P.L. 111-11, F, § 1503(b)(11)	P.L. 111-11, K, § 1803(g)	P.L. 111-11, L, § 1853(d)(7)	P.L. 111-11, O, § 1972(b)(5)

Access for Mineral Activities

P.L. 88-577, §§ 4(d)(2) & (3) P.L. 98-428, § 304	P.L. 95-237, § 4(g) P.L. 103-433, § 106	P.L. 96-312, § 5(d) P.L. 104-333, § 1022(a)(4)	P.L. 98-425, § 110 P.L. 106-456, § 3(a)

Access for Other Specific Activities

Commercial Activities			
P.L. 88-577, § 4(d)(6) P.L. 111-11, K, § 1803(j) & (k)	P.L. 96-312, § 7(a)(3) P.L. 111-11, M, § 1903(e)	P.L. 111-11, A, § 1001(c)	P.L. 111-11, F, § 1503(b)(6)
	Cemeteries		
P.L. 97-384, § 3	P.L. 101-633, § 9		

(Continued).

	Rights-of-Way		
P.L. 95-249, § 4	P.L. 98-425, § 101(a)(24)	P.L. 103-433, § 102(13)	
	Tribal Uses		
P.L. 107-282, § 206 P.L. 111-11, L, § 1853(d)(6)	P.L. 108-424, § 207 P.L. 111-11, O, § 1972(b)(7)	P.L. 109-362, § 4(k)	P.L. 109-432, § 327

Appendix B. Individual Statutes

This is a chronological list of statutes, or titles or subtitles within a statute, that designate wilderness areas, including the management provisions and any exceptions to general wilderness management for particular areas.[22] Be aware that, except for the Wilderness Act itself, these wilderness statutes are generally not codified; a collected set of the statutes can be found at http://www.wilderness.net/index.cfm?fuse=docmap.

P.L. 88-577—The Wilderness Act

§ 2—defines wilderness and identifies the purpose of wilderness designation as protecting "undeveloped Federal land retaining its primeval character and influence, without permanent improvements or human habitation, which is protected and managed so as to preserve its natural conditions."

§ 3(a)—designates "all areas within the national forests classified ... as 'wilderness', 'wild', or 'canoe' ... as wilderness areas," establishing 54 new USFS wilderness areas totaling 9,139,721 acres in 13 states.

§ 3(b)—directs the Secretary of Agriculture, "within ten years after the enactment" to review "as to its suitability or nonsuitablity for preservation as wilderness, each area ... classified ... as 'primitive.'" Recommendations from the President are to be presented within three years for at least 1/3 of the primitive areas and within seven years for at least 2/3 of the primitive areas. Directs management of the primitive areas to continue under existing rules and regulations.

§ 3(c)—directs the Secretary of the Interior to review "every roadless area of five thousand contiguous acres or more in the national parks, monuments and other units of the national park system and every such area of, and every roadless island within, the national wildlife refuges and game ranges." Recommendations from the President are to be presented within three years for at least 1/3 of the areas reviewed and within seven years for at least 2/3 of the areas reviewed.

§ 4(c)—"except as specifically provided for in this Act, and subject to existing private rights, there shall be no commercial enterprise and no permanent road within any wilderness area ... and, except as necessary to meet minimum requirements for the administration of the area ...

(including measures required in emergencies involving the health and safety of persons within the area), there shall be no temporary road, no use of motor vehicles, motorized equipment or motorboats, no landing of aircraft, no other form or mechanical transport, and no structure or installation within any such area."

§ 4(d)(1)—"the use of aircraft or motorboats, where these uses have already become established, may be permitted to continue subject to such restrictions as the Secretary ... deems desirable."

§ 4(d)(1)—"such measures may be taken as may be necessary in the control of fire, insects, and diseases, subject to such conditions as the Secretary deems desirable."

§ 4(d)(2)—"nothing in this Act shall prevent ... any activity, including prospecting, for the purpose of gathering information about mineral or other resources, if such activity is carried on in a manner compatible with the preservation of the wilderness environment."

§ 4(d)(3)—"notwithstanding any other provisions of this Act, until midnight December 31, 1983, the United States mining laws and all laws pertaining to mineral leasing shall, to the same extent as applicable prior to the effective date of this Act, extend to ... [each] 'wilderness area'; subject, however, to such reasonable regulations ... as may be prescribed ... consistent with the use of the land for mineral location and development. ... Mineral leases, permits, and licenses ... shall contain such reasonable stipulations as may be prescribed ... for the protection of the wilderness character of the land consistent with the [lease, permit, or license]. Subject to valid rights then existing, effective January 1, 1984, the minerals in ... wilderness areas are withdrawn from all forms of appropriation under the mining laws and from disposition under all laws pertaining to mineral leasing."

§ 4(d)(4)(1)—"the President may, within a specific area and in accordance with such regulations as he may deem desirable, authorize prospecting for water resources, the establishment and maintenance of reservoirs, water-conservation works, power projects, transmission lines, and other facilities needed in the public interest, including the road construction and maintenance essential to development and use thereof."

§ 4(d)(4)(2)—"the grazing of livestock, where established prior to the effective date of this Act, shall be permitted to continue subject to such reasonable regulations as are deemed necessary."

§ 4(d)(5)—"nothing in this Act shall preclude the continuance within the [Boundary Waters Canoe] area of any already established use of motorboats."

§ 4(d)(6)—"commercial services may be performed within the wilderness areas ... to the extent necessary for activities which are proper for realizing the recreational or other wilderness purposes."

§ 4(d)(7)—"nothing in this Act shall constitute an express or implied claim or denial on the part of the Federal Government as an exemption from State water laws."

§ 4(d)(8)—"nothing in this Act shall be construed as affecting the jurisdiction or responsibilities of the several States with respect to wildlife and fish."

§ 5(a)—"in any case where State-owned or privately owned land is completely surrounded by ... wilderness, such State or private owner shall be given such rights as may be necessary to assure adequate access ... or the ... land shall be exchanged for federally owned land in the same State of approximately equal value."

§ 5(b)—"in any case where valid mining claims or other valid occupancies are wholly within a designated ... wilderness area, the Secretary ... shall, by reasonable regulations consistent with the preservation of the area as wilderness, permit ingress and egress to such surrounded areas by means which have been or are being customarily enjoyed with respect to other such areas similarly situated."

§ 5(c)—"the Secretary is authorized to acquire privately owned land within the perimeter of any ... wilderness if (1) the owner concurs in such acquisition or (2) the acquisition is specifically authorized by Congress."

P.L. 90-271—San Raphael Wilderness

§ 1—designates one new USFS wilderness area with 143,000 acres in California. § 3—directs management in accordance with the provisions of the Wilderness Act.

P.L. 90-318—San Gabriel Wilderness

§ 1—designates one new USFS wilderness area with 36,000 acres in California.

§ 3—directs management in accordance with the provisions of the Wilderness Act.

P.L. 90-532—Great Swamp Wilderness

§ 1—designates one new FWS wilderness area with 3,750 acres in New Jersey.

§ 3—"except as necessary to meet minimum requirements ... (including measures required in emergencies involving the health and safety of persons within the area), there shall be no commercial enterprise, no temporary or permanent roads, no use of motor vehicles, motorized equipment or motorboats, no landing of aircraft, no other form of motorized transport, and no structure or installation within the area."

P.L. 90-544—North Cascades National Park, Recreation, and Wilderness Areas, Title VI

§ 601(a)—designates one new USFS wilderness area (with about 500,000 acres) in Washington. § 602—designates one USFS wilderness area addition (with about 10,000 acres) in Washington.

§ 604—requires a review of the wilderness potential of "the area within the North Cascades National Park" and recommendations to the President within two years.

P.L. 90-548—Mount Jefferson Wilderness

§ 1—designates one new USFS wilderness area with "approximately 100,000 acres" in Oregon. § 3—directs management in accordance with the provisions of the Wilderness Act.

P.L. 91-58—Ventana Wilderness

§ 1—designates one new USFS wilderness area with 98,000 acres in California.

§ 3—directs management in accordance with the provisions of the Wilderness Act.

P.L. 91-82—Desolation Wilderness

§ 1—designates one new USFS wilderness area with 63,000 acres in California.

§ 3—directs management in accordance with the provisions of the Wilderness Act. "Owners and operators of existing federally licensed hydroelectric facilities shall have the right of reasonable access to the areas for purposes of operating and maintaining such facilities in a manner that is consistent with past practices without prior approval of the Secretary."

P.L. 91-504—To designate wilderness areas

§ 1—designates 20 new FWS wilderness areas totaling 99,025 acres in 10 states.

§ 2—designates two new NPS wilderness areas totaling 93,503 acres in two states. § 3—designates one new USFS wilderness area with 7,000 acres in one state.

§ 5—directs management in accordance with the provisions of the Wilderness Act.

P.L. 92-230—Pine Mountain Wilderness

§ 1—designates one new USFS wilderness area with 19,700 acres in Arizona.

§ 3—directs management in accordance with the provisions of the Wilderness Act.

P.L. 92-241—Sycamore Canyon Wilderness

§ 1—designates one new USFS wilderness area with 48,500 acres in Arizona.

§ 3—directs management in accordance with the provisions of the Wilderness Act.

P.L. 92-364—Cedar Keys Wilderness

§ 1—designates one new FWS wilderness area in Florida.

§ 3—directs management in accordance with the provisions of the Wilderness Act.

P.L. 92-395—Scapegoat Wilderness

§ 1—designates one new USFS wilderness area with 240,000 acres in Montana.

§ 3—directs management in accordance with the provisions of the Wilderness Act.

P.L. 92-400—Sawtooth National Recreation Area and Wilderness

§ 1(b)—designates one new USFS wilderness area in Idaho.

§ 2(b)—directs management in accordance with the provisions of the Wilderness Act.

§ 5—requires a review of the wilderness suitability of "the undeveloped and unimproved portion" of the recreation area.

P.L. 92-476—Washakie Wilderness

§ 1—designates one new USFS wilderness area with 208,000 acres in Wyoming.

§ 5—designates a special management area; requires a study of the wilderness potential of that area and recommendations within five years.

P.L. 92-493—Lava Beds Wilderness

§ 1—designates one new NPS wilderness area with 28,460 acres in California.

§ 3—directs management in accordance with the provisions of the Wilderness Act.

P.L. 92-510—Lassen Volcanic Wilderness

§ 1—designates one new NPS wilderness area with 78,982 acres in California.

§ 3—directs management in accordance with the provisions of the Wilderness Act.

P.L. 92-521—Eagle Cap Wilderness Addition

§ 1—designates additions to one USFS wilderness area with 72,420 acres in Oregon. § 3—directs management in accordance with the provisions of the Wilderness Act.

§ 4—designates one USFS wilderness study area with 32,000 acres in Oregon, with recommendations within five years.

P.L. 92-528—Indian Peaks Wilderness Study

§ 1(a)—designates one USFS wilderness study area with 71,000 acres in Colorado.

P.L. 93-429—Okefenokee Wilderness

§ 1—designates one new FWS wilderness area with 343,850 acres in Georgia.

§ 2—"subject to such restrictions as the Secretary deems necessary for public safety and to protect flora and fauna, (1) the use of powered watercraft, propelled by motors of ten or less horsepower, will be permitted, (2) watercraft trails ... will be maintained." Authorizes access to the watercraft trails from four specific entry points.

§ 3—authorizes fishing, and allows zones and periods of no fishing "for reasons of public safety, administration, fish and wildlife management, or public use and enjoyment."

§ 5—directs management in accordance with the provisions of the Wilderness Act.

P.L. 93-550—Farallon Wilderness

§ 101—designates one new FWS wilderness area with 141 acres in California.

§ 102—directs management in accordance with the provisions of the Wilderness Act.

P.L. 93-622—The Eastern Wilderness Act

§ 3—designates 16 new USFS wilderness areas totaling 206,988 acres in 13 states.

§ 4(b)—designates 17 USFS wilderness study areas totaling 125,000 acres in nine states.

§ 4(c)—requires a review of the study areas and recommendations to Congress within five years.

§ 6(a)—directs management of the wilderness areas in accordance with the provisions of the Wilderness Act. Directs management of the study areas "to maintain their presently existing wilderness character and potential ... until Congress has determined otherwise, except that such management requirement shall in no case extend beyond the expiration of the third succeeding Congress from the date of submission to Congress of the President's recommendations."

§ 6(b)(1)—"the Secretary may acquire by purchase with donated or appropriated funds, by gift, exchange, condemnation, or otherwise, such lands, waters, or interests therein as he determines necessary or desirable."

§ 6(b)(3)—"the authority of the Secretary ... to condemn any private land or interest therein ... shall not be invoked so long as the owner or owners of such land or interest holds and uses it in the same manner and for those purposes for which such land or interest was held on the date of the designation ...: *Provided, however*, That the Secretary ... may acquire such land or interest without consent ... whenever he finds such use to be incompatible with the management of such area as wilderness and the owner or owners manifest unwillingness, and subsequently fail, to promptly discontinue such incompatible use."

§ 6(b)(4)—requires 60 days notice of any ownership transfer for inholdings.

§ 6(b)(5)—requires 60 days notice for inholdings of "any change in the use ... which will result in any significant new construction or disturbance of land surface or flora or will require the use of motor vehicles and other forms of mechanized transport or motorized equipment (except as otherwise authorized by law for ingress or egress or for existing agricultural activities begun before the date of the designation other than timber cutting)."

§ 6(b)(7)—as a condition of transfer, owners may "retain for themselves and their successors or assigns a right of use and occupancy of the property for such noncommercial residential purpose or agricultural activity for twenty-five years."

§ 6(b)(8)—"a right of use or occupancy retained ... may be terminated with respect to the entire property by the Secretary ... upon his determination that the property or any portion thereof has ceased to be used for such noncommercial residential purpose or agricultural activity."

P.L. 93-632—To designate wilderness areas

§ 1—designates 13 new FWS wilderness areas totaling 116,006 acres in 10 states.

§ 2—designates four new USFS wilderness areas totaling 604,500 acres in three states. § 5—directs management in accordance with the provisions of the Wilderness Act.

P.L. 94-146—Flat Tops Wilderness

§ 1—designates one new USFS wilderness area with 235,230 acres in Colorado.

§ 3—directs management in accordance with the provisions of the Wilderness Act.

P.L. 94-199—Hells Canyon National Recreation Area & Wilderness

§ 2(a)—designates one new USFS wilderness area in Idaho and Oregon.

§ 2(b)—directs management in accordance with the provisions of the Wilderness Act.

§ 8(d)—requires a review of the wilderness suitability of three specified areas and recommendations within five years. Directs management of the areas "so as not to preclude their possible future designation by the Congress as wilderness."

P.L. 94-268—Wilderness Boundary Adjustment

§ 1—reduces one USFS wilderness area with 6,500 acres in Vermont (designated in P.L. 93-622, § 3(a)(10)) by deleting the area (§ 1(a)) and then designating a new area of 3,775 acres (§ 1(b)(2)).

§ 1(b)(1)—redesignates one USFS wilderness area in Georgia and Tennessee, with the same acreage as enacted in § 3(b) of P.L. 93-622.

P.L. 94-352—Eagles Nest Wilderness

§ 1—designates one new USFS wilderness area with 133,900 acres in Colorado.

§ 3—directs management in accordance with the provisions of the Wilderness Act.

P.L. 94-357—The Alpine Lakes Area Management Act of 1976

§ 3(a)—designates one new USFS wilderness area and one USFS "Intended Wilderness" in Washington.

§ 3(c)—directs management the wilderness areas in accordance with the provisions of the Wilderness Act.

§ 3(d) & (e)—adds the "Intended Wilderness" after acquisition of "the adjacent non-Federal lands, interests or other property."

P.L. 94-544—Point Reyes Wilderness

§ 1—designates one new NPS wilderness area with 25,370 acres and one potential NPS wilderness addition of 8,003 acres in California.

§ 3—directs management of the wilderness area in accordance with the provisions of the Wilderness Act.

P.L. 94-557—To designate wilderness areas

§ 1—designates 16 new FWS wilderness areas totaling 155,156 acres in 11 states.

§ 2—designates three new USFS wilderness areas totaling 232,415 acres in three states.

§ 3(b)—designates eight USFS wilderness study areas totaling 587,364 acres in three states. Requires a review and recommendations to Congress within 19 months (one area), two years (three areas), or five years (four areas).

§ 3(d)—directs management of the study areas "subject to valid existing rights, ... until Congress has determined otherwise, ... to maintain their presently existing wilderness character and potential ..., except that such management requirement shall in no case extend beyond a period of four years from the date of submission to Congress of the President's recommendation. ... Already established uses may be permitted to continue, subject to such restrictions as the Secretary of Agriculture deems desirable, in the manner and degree which the same was being conducted on the date of enactment."

§ 6—directs management of the wilderness areas in accordance with the provisions of the Wilderness Act.

P.L. 94-567—To Designate Wilderness Areas

§ 1—designates 13 new NPS wilderness areas totaling 919,268 acres in eight states and eight potential NPS wilderness area additions totaling 53,506 acres in six states.[23]

§ 3—adds the potential wilderness "upon publication in the Federal Register of a notice by the Secretary of the Interior that all uses thereon prohibited by the Wilderness Act have ceased."

§ 6—directs management of wilderness areas in accordance with the provisions of the Wilderness Act.

§ 8—deletes 6,497 acres from one USFS wilderness area in Wyoming; designates it as a primitive area and requires a review of its wilderness potential.

P.L. 94-579—The Federal Land Policy and Management Act of 1976

§ 603(a)—requires a review of the wilderness potential of "those roadless areas of five thousand acres or more and roadless islands of the public lands" (defined in § 103(e) as lands administered by the BLM).

§ 603(c)—directs management of those identified wilderness study areas (WSAs) "until Congress has determined otherwise, ... in a manner so as not to impair the suitability of such areas for preservation as wilderness, subject, however, to the continuation of existing mining and grazing uses and mineral leasing in the manner and degree in which the same were being conducted" prior to their identification as WSAs.

P.L. 95-150—The Montana Wilderness Study Act of 1977

§ 2(a)—designates nine USFS wilderness study areas totaling 973,000 acres in Montana. § 2(b)—requires a review and recommendations to Congress within seven years.

§ 3(a)—directs management of the study areas "subject to existing private rights, ... until Congress determines otherwise, ... to maintain their presently existing wilderness character."

§ 3(b)—states that the statute is to have no effect on state jurisdiction over or responsibilities for fish and wildlife.

P.L. 95-237—The Endangered American Wilderness Act of 1978

§ 2—designates 10 new USFS wilderness areas and one USFS wilderness area addition totaling 710,267 acres in seven states.

§ 2(c)—"to guarantee the continued viability of the Santa Lucia watershed and to insure the continued health and safety of the communities serviced by such watershed, the management plan for the Santa Lucia area to be prepared following designation as wilderness shall authorize the Forest Service to take whatever appropriate actions are necessary for fire prevention and watershed protection including, but not limited to, acceptable fire presuppression and fire suppression measures and techniques."

§ 2(e)—establishes "no right, or claim of right, to the diversion and use of the waters of Hunter Creek, the Fryingpan or Roaring Fork Rivers, or any tributaries of said creed or rivers, by the Fryingpan-Arkansas Project ... shall be prejudiced, expanded, diminished, altered, or affected by this Act. Nothing in this Act shall be construed to expand, abate, impair, impede, or interfere with the construction, maintenance, or repair of said Fryingpan-Arkansas Project facilities, not the operation thereof."

§ 2(i)—authorizes "whatever sanitary facilities are necessary (including but not limited to vault toilets, which may require service by helicopter) to insure the continued health and safety of the communities serviced by the Lone Peak watershed; furthermore, nothing in this Act shall be construed to limit motorized access and road maintenance by local municipalities for those minimum maintenance activities necessary to guarantee the continued viability of whatsoever watershed facilities currently exist, or which may be necessary in the future to prevent the degradation of the water supply in the Lone Peak area."

§ 3—**Oregon Omnibus Wilderness Act of 1978**.

Designates two new USFS wilderness areas and three USFS wilderness area additions totaling 387,100 acres in Oregon.

§ 4—**Gospel-Hump Area**.

§ 4(a)(1)—designates one new USFS wilderness area with 206,000 acres in Idaho.

§ 4(g)—provides that "nothing in this Act shall prevent within the Gospel-Hump Wilderness Area any activity, including prospecting, for the purpose of gathering information about mineral or other resources, if such activity is carried on in a manner compatible with the preservation of the wilderness environment."

§ 5—directs management in accordance with the provisions of the Wilderness Act, subject to valid existing rights.

P.L. 95-249—Absaroka-Beartooth Wilderness

§ 1—designates one new USFS wilderness area with 904,500 acres in Montana.

§ 3—directs management in accordance with the provisions of the Wilderness Act.

§ 4—states that the statute is to have no effect on any claim by two counties to a specific right-of-way.

P.L. 95-450—The Indian Peaks Wilderness Area, the Arapaho National Recreation Area and the Oregon Islands Wilderness Area Act

§ 3(a)—designates one new USFS wilderness area with 70,000 acres in Colorado.

§ 3(b)—directs management in accordance with the provisions of the Wilderness Act. § 12(a)—designates one new FWS wilderness areas with 459 acres in Oregon.

§ 12(b)—directs management in accordance with the provisions of the Wilderness Act.

P.L. 95-494—Wisconsin Wilderness

§ 1—designates two new USFS wilderness areas totaling 13,201 acres in Wisconsin.

§ 3—directs management in accordance with the provisions of the Wilderness Act.

P.L. 95-495—Boundary Waters Canoe Area Wilderness

§ 3—designates one new USFS wilderness area with 1,075,500 acres in Minnesota.

§ 4(a)—directs management in accordance with the provisions of the Wilderness Act.

§§ 4(c) & (d)—identify lakes where the use of motors up to the specified horsepower are permitted and allow mechanized portages in specified areas.

§ 4(e)—allows use of snowmobiles, within limited sizes, in specified areas.

§ 4(f)—specifies entry point quotas for motorboats on the authorized lands.

§ 4(g)—permits continued motor vehicle assistance in certain portages until January 1, 1984, and thereafter if "there is no feasible nonmotorized means of transporting boats across the portages."

§ 4(h)—limits all motorized uses to "those types of snowmobiles, motorboats, and vehicles which have been in regular use" prior to enactment.

§ 4(i)—prohibits motorized use except as specified and for emergencies and administration.

§ 8—incorporates the "provisions of Executive Order 10092" concerning an existing airspace reservation.

§ 11(a)—withdraws the area from mining and mineral exploration.

§ 13—permits the maintenance of a particular dam and of "other existing water control structures only where ... necessary to protect wilderness values or public safety."

§ 14—directs that state jurisdiction over or responsibilities for fish and wildlife are unaffected.

§ 15—states that the statute is to have no effect on state jurisdiction over or responsibilities for waters, "except to the extent that the exercise of such jurisdiction is less stringent that the Secretary's regulations." Requires that "any regulations ... be complementary to, and not in derogation of regulations issued by the United States Coast Guard."

§ 16(a)—requires cooperation with the state "and any political subdivisions thereof." § 17—states that the statute is to have no effect on any treaty.

§ 18—authorizes and directs the Secretary "to expedite and intensify the program of dispersed outdoor recreation development, ... [including] additional snowmobile trails, ... remote campsites, ... lack access and parking facilities, [and] ... a system of new hiking, backpacking, and cross-country ski trails."

§ 19(b)—authorizes grants for "educational and technical assistance to businesses and communities adjacent ... to improve economic opportunities for tourism and recreation-related businesses in a manner which is complementary to the management of the wilderness."

P.L. 95-546—Great Bear Wilderness

§ 1—designates one new USFS wilderness area with 285,771 acres and one USFS wilderness addition with 60,000 acres in Montana.

§ 3—directs management in accordance with the provisions of the Wilderness Act.

P.L. 95-625—The National Parks and Recreation Act of 1978, Title IV

§ 401—designates eight new NPS wilderness areas totaling 1,854,395 acres in eight states and six potential NPS wilderness area additions totaling 94,139 acres in six states.

§ 403—adds the potential wilderness "upon publication in the Federal Register of a notice by the Secretary that all uses thereon prohibited by the Wilderness Act have ceased." Directs management "insofar as practicable as wilderness."

§ 404—directs management in accordance with the provisions of the Wilderness Act.

P.L. 96-248—Sandia Mountain Wilderness Addition

§ 1—expands one USFS wilderness area in New Mexico and withdraws the additional area from the mining and mineral leasing laws.

P.L. 96-312—The Central Idaho Wilderness Act of 1980

§ 3—designates one new USFS wilderness area with 2,239,000 acres in Idaho.

§ 4—designates one USFS wilderness area addition with 105,600 acres in Idaho.

§ 5(a)—requires "a comprehensive wilderness management plan" within three years.

§ 5(c)—directs management in accordance with the provisions of the Wilderness Act, subject to valid existing rights.

§ 5(d)(1)—establishes within the wilderness area the "Special Mining Management Zone—Clear Creek," and directs that "all prospecting and exploration for, and development or mining of cobalt and associated minerals shall be considered a dominant use of such lands."

§ 5(d)(1)(A)—"all mining locations and associated access roads shall be held and used solely for mining and minerals processing operations and uses reasonably incident thereto," except that the state may use the access roads "to facilitate the management of the bighorn sheep" in the zone.

§ 5(d)(1)(B)—directs that mining law patents convey title only to the minerals; provides "the right to cut and use as much of the mature timber therefrom as may be needed in the extraction, removal and beneficiation of the mineral deposits, if such needed timber is not otherwise reasonably available, and if such timber is cut under sound principles of forest management;" and provides the patentee "the right to use as much of the surface as reasonably necessary for the mining, removal, extraction, of beneficiation of the mineral deposits."

§ 5(d)(1)(C)—requires "all reasonable measures to see that the mining or processing of cobalt ... does not significantly impair the overall habitat of bighorn sheep."

§ 7(a)(1)—allows "the landing of aircraft, where this use has become established prior to the date of enactment ... to continue subject to such restrictions as the Secretary deems desirable." Prohibits the Secretary from acting to "permanently close or render unserviceable any aircraft landing strip in regular use ... for reasons other than extreme danger to aircraft."

§ 7(a)(2)—allows continued livestock grazing where "established before the date of enactment ... subject to such reasonable regulations as the Secretary deems necessary" in accordance with the Wilderness Act.

§ 7(a)(3)—authorizes "commercial services ... to the extent necessary for activities which are proper for realizing the recreational or other wilderness purposes."

§ 7(a)(4)—authorizes "future construction and maintenance of small hydroelectric generators, domestic water facilities, and related facilities" in two drainages.

§ 7(b)—provides that nothing "shall constitute an express or implied claim or denial on the part of the Federal Government as to exemption from State water laws."

§ 7(c)—directs that state jurisdiction over and responsibilities for fish and wildlife are unaffected.

P.L. 96-476—The Rattlesnake National Recreation Area and Wilderness Act of 1980

§ 2(a)—designates one new USFS wilderness area with 33,000 acres in Montana.

§ 2(b)—directs management in accordance with the provisions of the Wilderness Act, subject to valid existing rights.

P.L. 96-487—The Alaska National Interest Lands Conservation Act

§ 701—designates eight new NPS wilderness areas totaling 32,355,000 acres in Alaska. § 702—designates eight new FWS wilderness areas totaling 18,560,000 acres in Alaska. § 703(a)—designates 14 new USFS wilderness areas totaling 5,761,499 acres in Alaska. § 703(b)—allows continued use of existing mechanized portage equipment at one site.

§ 704—designates one USFS wilderness study area in Alaska, with recommendations within three years.

§ 707—directs management of the wilderness areas in accordance with the Wilderness Act. § 708—provides statewide USFS release language, except for the area identified in § 704.

P.L. 96-550—New Mexico Wilderness

§ 102(a)—designates eight new USFS wilderness areas and four USFS wilderness area additions totaling 609,060 acres in New Mexico.

§ 102(a)(5)—states that the statute is to have no effect on additional fencing or grazing levels in one unit.

§ 102(a)(9)—states that the statute is to have no effect on "the management of, or rules, regulations and law applying to the Santa Fe Municipal Watershed" in one unit.

§ 102(d)—directs management in accordance with the provisions of the Wilderness Act, subject to valid existing rights.

§ 103—designates six USFS wilderness study areas with 117,530 acres in New Mexico, with recommendations by January 1, 1986. Directs management of the areas "until Congress determines otherwise ... so as to maintain their presently existing wilderness character ... [but] current levels of motorized and other uses shall be permitted to continue such to ... reasonable rules and regulations."

§ 104—provides statewide USFS release language, except for the area identified in § 103. § 105—states that the statute creates no buffer zone.

P.L. 96-560—Colorado Wilderness

§ 102(a)—designates 14 new USFS wilderness areas and six USFS wilderness area additions totaling 1,423,130 acres in Colorado.

§ 102(a)(5)—establishes "no right, or claim of right, to the diversion and use of existing conditional water rights for the Homestake Water Development project by the cities of Aurora and Colorado Springs shall be prejudiced, expanded, diminished, altered, or affected by this Act. Nothing in this Act shall be construed to expand, abate, impair, impede, or interfere with the construction, maintenance or repair of said project, not the operation thereof."

§ 102(a)(17)—allows "motorized access and the use of motorized equipment used for the periodic maintenance and repair of the McGuire Water Transmission Line ditch."

§ 103—designates one new USFS wilderness area with 10,700 acres in South Dakota; "the provisions of the Act designating the Custer State Park Sanctuary ... shall also apply ... to the extent they are not inconsistent with the provisions of the Wilderness Act."

§ 104(b)—directs management in accordance with the provisions of the Wilderness Act, subject to valid existing rights.

§ 105(a)—designates nine USFS wilderness study areas totaling 477,400 acres in Colorado, with recommendations within three years.

§ 105(b)—designates one joint USFS-BLM study of 14,000 acres in Colorado, with recommendations within three years. The study must consider a range of values, several possible land designations (including wilderness), the effect of various options on local and national economic activities, and access to the area.

§ 105(c)—directs management of the study areas "to maintain their presently existing wilderness character and potential" until Congress determines otherwise, but "grazing of livestock and oil, gas, or mineral exploration and development activities" shall be administered under laws generally applicable to USFS lands.

§ 106(a)—designates one additional USFS wilderness study area with 5,500 acres in Colorado, with recommendations within three years.

§ 106(b)—directs management of the study areas "to maintain their presently existing wilderness character and potential" for up to "two years from the date of submission" of the recommendation, with "oil, gas, or mineral exploration and development operations" permitted under the terms of the Wilderness Act.

§ 107—provides statewide USFS release language, except for areas identified in §§ 105 & 106.

§ 108—"declares that ... [for] livestock grazing in National Forest wilderness areas, the provisions of the Wilderness Act shall be ... administered in accordance with the guidelines contained" in H.Rept. 96-617.

§ 109—requires a review of "all practices, policies, and regulations ... regarding disease or insect outbreaks, forest fires, and the use of modern suppression methods and equipment ... to insure that—(a) such policies, practices, and regulations fully conform with and implement the intent of Congress" as expressed in the Wilderness Act, "and (b) policies, practices, and regulations are developed that will allow timely and efficient fire, insect, and disease control, to provide, to the extent reasonably practicable, adequate protection of adjacent [lands]."

§ 110—states that the statute creates no buffer zone.

§ 201—designates nine new USFS wilderness areas with 50,070 acres in three states.

§ 203—directs management in accordance with the provisions of the Wilderness Act, subject to valid existing rights.

H. Rept. 96-617 to Accompany H.R. 5487 (P.L. 96-560)— Grazing in National Forest Wilderness Areas

Pages 10-13 of the report express the Committee's concern that the intent of the Wilderness Act language authorizing continued livestock grazing in wilderness was not being implemented appropriately. The Committee "declined to amend section 4(d)(4)(2) of the Wilderness Act, opting instead for a reaffirmation of the ... language in section 5 of H.R. 5487 and for the following nationwide guidelines and specific statements of legislative policy." The report includes five guidelines:

1) Do not curtail or "phase-out" grazing simply because the area has been designated wilderness.
2) Allow maintenance of supporting facilities, including with motorized equipment when necessary.
3) Require "natural materials" to replace or reconstruct deteriorated facilities only if it would impose unreasonable additional costs.
4) Allow replacements or new improvements if primarily for resource protection.
5) Allow use of motorized equipment for emergencies.

P.L. 96-585—Otis Pike Fire Island High Dune Wilderness

§ 1(a)—designates one new NPS wilderness area with 1,363 acres in New York.

§ 1(d)—provides that the designation "shall not preclude the repair of breaches that occur in the wilderness area, in order to prevent loss of life, flooding, and other severe economic and physical damage to the Great South Bay and surrounding areas."

P.L. 97-211—Florida Keys Wilderness Addition

§ 1—adds 73 acres to an existing FWS wilderness in Florida after the lands have been added to the National Wildlife Refuge System.

§ 2—deletes 25 acres (one island) from the FWS wilderness in Florida.

§ 4—directs management in accordance with the Wilderness Act, subject to valid existing rights.

P.L. 97-250—Cumberland Island Wilderness

§ 2(a)—designates one new NPS wilderness area with 8,840 acres in Georgia. Also designates potential NPS wilderness with 11,718 acres, to be added (in whole or piecemeal) when the Secretary publishes a *Federal Register* notice that uses prohibited by the Wilderness Act have ceased.

§ 2(c)—directs management in accordance with the Wilderness Act, subject to valid existing rights.

P.L. 97-384—Charles C. Deam Wilderness

§ 1—designates one new USFS wilderness area with 12,953 acres in Indiana.

§ 2—directs management in accordance with the Wilderness Act, subject to valid existing rights.

§ 3—states that the statute is to have no effect on "the right of public access to cemeteries" and access to privately owned lands surrounded by wilderness, in accordance with the Wilderness Act.

§ 4—provides statewide USFS release language.

P.L. 97-407—Paddy Creek Wilderness

§ 2—designates one new USFS wilderness area with 6,888 acres in Missouri.

§ 4—directs management in accordance with the Wilderness Act, subject to valid existing rights. § 5—provides statewide USFS release language.

P.L. 97-411—Cheaha Wilderness

§ 2(a)—designates one new USFS wilderness area with 6,780 acres in Alabama.

§ 2(b)—directs management in accordance with the Wilderness Act, subject to valid existing rights.

P.L. 97-466—Monongahela Wilderness

§ 1—designates three new USFS wilderness areas totaling 47,800 acres in West Virginia.

§ 3—directs management in accordance with the Wilderness Act, subject to valid existing rights.

§ 4—requires acquisition of "all nonfederally owned coal deposits and other mineral interests and rights" in exchange for credit that can be used for federal coal leases elsewhere.

§ 5—provides statewide USFS release language.

P.L. 98-140—The Lee Metcalf Wilderness and Management Act of 1983

§ 2(a)—designates one new USFS/BLM wilderness area with 259,000 acres in Montana.

§ 2(b)—directs management in accordance with the Wilderness Act, subject to valid existing rights.

§ 3(a)-(c)—provides USFS and BLM release language for specified areas.

§ 3(d)—revises the boundary of one USFS wilderness area in Montana, deleting 67 acres. § 3(e)—revises the boundary of one FWS wilderness area in Montana, deleting 28 acres.

P.L. 98-289—Irish Wilderness

§ 2(a)—designates one new USFS wilderness area with 16,500 acres in Missouri.

§ 2(b)—directs management in accordance with the Wilderness Act, subject to valid existing rights.

P.L. 98-321—The Wisconsin Wilderness Act of 1984

§ 2—designates two new USFS wilderness areas (one with three separate units) totaling 24,339 acres in Wisconsin.

§ 4—directs management in accordance with the Wilderness Act, subject to valid existing rights. § 5—provides statewide USFS release language.

P.L. 98-322—The Vermont Wilderness Act of 1984

§ 102—designates four new USFS wilderness areas and one USFS wilderness area addition totaling 41,260 acres in Vermont.

§ 104(a)—directs management in accordance with the Wilderness Act, subject to valid existing rights.

§ 104(b)—directs that state jurisdiction over and responsibilities for fish and wildlife are unaffected.

§ 104(c)—authorizes maintenance of two specific trails, associated trails, and "related structures." § 105—provides statewide USFS release language.

P.L. 98-323—The New Hampshire Wilderness Act of 1984

§ 101—designates three new USFS wilderness areas and one USFS wilderness area addition totaling 77,000 acres in New Hampshire.

§ 103—directs management in accordance with the Wilderness Act, subject to valid existing rights.

§ 104—provides statewide USFS release language, except for the area specified in § 104(e).

P.L. 98-324—The North Carolina Wilderness Act of 1984

§ 2—designates nine new USFS wilderness areas and two USFS wilderness area additions totaling 68,750 acres in North Carolina.

§ 4—directs management in accordance with the Wilderness Act, subject to valid existing rights. § 5—provides statewide USFS release language, except for the area specified in § 6(a).

§ 6(a)—designates four new USFS wilderness study areas and one USFS wilderness study area addition totaling 25,816 acres in North Carolina.

§ 6(b)—requires recommendations within three years.

§ 6(c)—directs management, subject to valid existing rights, "so as to maintain their presently existing wilderness character" until Congress determines otherwise.

P.L. 98-328—The Oregon Wilderness Act of 1984

§ 3—designates 22 new USFS wilderness areas and seven USFS wilderness area additions totaling 788,700 acres in Oregon.

§ 4(f)—designates one new USFS wilderness area and one USFS wilderness area addition totaling 70,800 acres in Oregon.

§ 5(b)—directs management in accordance with the Wilderness Act, subject to valid existing rights.

§ 6—states that the statute creates no buffer zone. § 7—provides statewide USFS release language.

P.L. 98-339—The Washington State Wilderness Act of 1984

§ 3—designates 18 new USFS wilderness areas and two USFS wilderness area additions totaling 898,790 acres in Washington.

§ 3(7)—revises the boundary of one USFS wilderness area in Washington, effectively deleting 800 acres.

§ 4(b)—directs management in accordance with the Wilderness Act, subject to valid existing rights.

§ 5—provides statewide USFS release language.

§ 6(a)—designates one new BLM wilderness area with 7,140 acres in Washington.

§ 6(b)—directs management in accordance with the Wilderness Act, subject to valid existing rights.

§ 9—states that the statute creates no buffer zone.

P.L. 98-406—The Arizona Wilderness Act of 1984

§ 101(a)—designates 26 new USFS wilderness areas and four USFS wilderness area additions totaling 658,540 acres in Arizona.

§ 101(a)(13)—states that the statute is to have no effect on "the installation and maintenance of hydrologic, meteorologic, or telecommunications facilities ... or limited motorized access to such facilities when nonmotorized access means are not reasonably available or when time is of the essence, subject to such conditions as the Secretary deems desirable, where such facilities and access are essential to flood warning, flood control, and water reservoir operation purposes."

§ 101(a)(23)—"the governmental agency having jurisdictional authority may authorize limited access to the area, for private and administrative purposes," along a specified route.

§ 101(b)—directs management in accordance with the Wilderness Act, subject to valid existing rights.

§ 101(d)—states that the statute creates no buffer zone.

§ 101(e)(1)—establishes no "express or implied claim or denial on the part of the Federal Government as to exemption from Arizona State water laws."

§ 101(e)(2)—directs that state jurisdiction over and responsibilities for fish and wildlife are unaffected.

§ 101(f)(1)—allows continued livestock grazing "where established before the date of enactment" in accordance with the Wilderness Act and § 108 of P.L. 96-560.

§ 101(f)(2)—requires a review of all grazing management "to insure that such policies, practices, and regulations fully conform with and implement the intent of Congress."

§ 102(a)—designates three USFS wilderness study areas totaling 67,930 acres in Arizona, with recommendations from the President by January 1, 1986.

§ 103—provides statewide USFS release language, except for the areas specified in § 102. § 202—designates one new BLM wilderness area with 6,670 acres in Arizona.

§ 203—directs management in accordance with the Wilderness Act, subject to valid existing rights.

§ 301(a)—designates nine new BLM and USFS wilderness areas totaling 397,300 acres in Arizona.

§ 302(a)—directs management in accordance with the Wilderness Act, subject to valid existing rights.

§ 302(b)—allows continued livestock grazing "where established before the date of enactment ... subject to such reasonable regulations, policies, and practices as the Secretary concerns deems necessary, as long as such regulations, policies, and practices conform with and implement the intent of Congress regarding grazing in such areas as ... expressed in the Wilderness Act."

P.L. 98-425—The California Wilderness Act of 1984

§ 101(a)—designates 24 new USFS wilderness areas and 15 USFS wilderness area additions totaling 1,297,930 acres in California.

§ 101(a)(2)—allows "continued motorized access to those previously existing facilities ... [for] permitted livestock grazing activities ... in the same manner and degree in which such access was occurring as of the date of enactment."

§ 101(a)(6)—modifies the boundary of one wilderness area, transferring land to another area and directing the fire road between them "to be closed to all motorized vehicles except ... for administrative purposes."

§ 101(a)(7)—allows "nonmotorized dispersed recreation to continue at a level not less than the level of use which occurred during calendar year 1979."

§ 101(a)(13)—allows "portions of the existing primitive road between the two wilderness areas ... be relocated for environmental protection or for other reasons," with boundary adjustments to the wilderness areas, as needed, "effective upon publication of a notice of such relocation in the Federal Register."

§ 101(a)(24)—allows a right-of-way, pursuant to an application within 10 years of enactment, and authorizes construction of a transmission line. If the power transmission line is constructed, "the corridor shall cease to be a part of" the wilderness, and the Secretary "shall publish a notice thereof in the Federal Register."

§ 101(a)(25)—states that nothing "shall be construed to prejudice, alter, or affect in any way, any rights or claims of right to the diversion and use of waters from the North Fork of the San Joaquin River, or in any way interfere with the construction, maintenance, repair, or operation of a hydroelectric power project." Authorizes "continued motorized access to those previously existing facilities ... related to permitted livestock grazing ... [and] operation and maintenance of the existing cabin located in" the area "in the same manner and degree in which such access and operation and maintenance ... were occurring as of the date of enactment."

§ 102(a)—designates three USFS planning areas (i.e., wilderness study areas) totaling 98,200 acres in California, with recommendations within three years.

§ 102(b)—directs management the planning areas "so as to maintain their presently existing wilderness character and potential" for four years, subject to valid existing rights.

§ 103(a)—directs management USFS wilderness areas in accordance with the Wilderness Act, subject to valid existing rights.

§ 103(b)(1)—allows continued livestock grazing "where established before the date of enactment ... subject to such reasonable regulations, policies, and practices as the Secretary deems necessary, as long as such regulations, policies, and practices fully conform with and implement the intent of Congress regarding grazing in such areas as ... expressed in the Wilderness Act."

§ 103(b)(2)—authorizes, "as provided in ... the Wilderness Act , ... such measures as are necessary in the control of fire, insects, and diseases, subject to such conditions as [the Secretary] deems desirable."

§ 103(b)(3)—directs management wilderness areas, "as provided in ... the Wilderness Act ... to preserve their wilderness character and to devote them to the public purposes of recreational, scenic, scientific, educational, conservation, and historical use."

§ 103(c)—requires "negotiations to acquire by exchange all or part of any privately owned lands within the national forest wilderness areas."

§ 106—designates two new NPS wilderness areas totaling 1,414,580 acres and two potential NPS wilderness additions totaling 3,650 acres in California.

§ 108—adds potential wilderness "upon publication in the Federal Register of a notice ... that all uses prohibited by the Wilderness Act have ceased." Directs management "insofar as practicable as wilderness."

§ 109—directs management NPS wilderness areas in accordance with the Wilderness Act.

§ 110—directs "mineral prospecting, exploration, development, and mining of cobalt and associated minerals" in one USFS roadless area proceed under the U.S. mining laws governing nonwilderness lands.

§ 111—provides statewide USFS release language, except for 65 areas listed in § 111(e), pending wilderness recommendations for those areas developed during of preparation of land management plans.

P.L. 98-428—The Utah Wilderness Act of 1984

§ 102—designates 12 new USFS wilderness areas totaling 749,550 acres in Utah.

§ 103(b)—directs management in accordance with the Wilderness Act, subject to valid existing rights.

§ 201—provides statewide USFS release language.

§ 301(a)—allows continued livestock grazing, in accordance with the Wilderness Act and § 108 of P.L. 96-560.

§ 301(b)—requires a review of all grazing management "to insure that such policies, practices, and regulations fully conform with and implement the intent of Congress."

§ 302(a)—establishes no "express or implied claim or denial on the part of the Federal Government as to exemption from Utah water laws."

§ 302(b)—authorizes "whatever sanitary facilities are necessary, including but not limited to vault toilets which may require service by helicopter, to insure the continued health and safety of the communities serviced by the watersheds" for 10 of the wilderness areas. Also allows "motorized access and road maintenance by local municipalities for those minimum maintenance activities necessary to guarantee the continued viability of whatsoever watershed facilities currently exist or which may be necessary in the future to prevent the degradation of the water supply ... subject to such reasonable regulations as are deemed necessary."

§ 303—states that the statute creates no buffer zone.

§ 304—requires continued assessments of the mineral potential of the designated areas.

§ 305—authorizes "the installation and maintenance of hydrologic, meteorologic, climatological, or telecommunication facilities ... or limited motorized access to such facilities when nonmotorized access means are not reasonably available or when time is of the essence, subject to

such conditions as [are deemed] desirable, where such facilities or access are essential to flood warning, flood control and water reservoir operation."

§ 306(a)(1)—withdraws areas from mining laws and mineral and geothermal leasing laws, except for one area, where competitive CO_2 leases are authorized for bid within five years. Subsection (b) establishes conditions and requirements in the event of development of such leases.

P.L. 98-430—The Florida Wilderness Act of 1983

§ 1—designates six new USFS wilderness areas and one USFS wilderness area addition totaling 49,150 acres in Florida.

§ 1(4)—allows existing motorboat use on one creek.

§ 2(a)—designates two USFS wilderness study areas with 10,900 acres in Florida.

§ 2(b)—directs management of the study areas "so as to maintain their presently existing wilderness character and potential" until Congress determines otherwise, subject to valid existing rights.

§ 4—directs management of the wilderness areas in accordance with the Wilderness Act, subject to valid existing rights.

§ 6—provides statewide USFS release language.

P.L. 98-508—The Arkansas Wilderness Act of 1984

§ 3—designates eight new USFS wilderness areas and one USFS wilderness area addition totaling 91,103 acres in Arkansas.

§ 4—provides statewide USFS release language.

§ 6—directs management in accordance with the Wilderness Act, subject to valid existing rights. § 7—states that the statute creates no buffer zone.

P.L. 98-514—The Georgia Wilderness Act of 1984

§ 2—designates two USFS wilderness area additions totaling 14,439 acres in Georgia.

§ 4—directs management in accordance with the Wilderness Act, subject to valid existing rights.

§ 5—provides statewide USFS release language, except for 10 areas totaling 118,000 acres in Georgia.

P.L. 98-515—The Mississippi National Forest Wilderness Act of 1984

§ 2—designates two USFS wilderness areas totaling 5,440 acres in Mississippi.

§ 4—directs management in accordance with the Wilderness Act, subject to valid existing rights. § 5—provides statewide USFS release language.

P.L. 98-550—The Wyoming Wilderness Act of 1984

§ 201(a)—designates eight new USFS wilderness areas and five USFS wilderness area additions totaling 884,129 acres in Wyoming.

§ 201(a)(11)—authorizes "occasional motorized access for administrative purposes and related activities ... for habitat management, trapping, transporting and proper management of the area's bighorn sheep population."

§ 201(c)—provides that four designated areas shall not affect "any present or future water rights for, and ... construction, operation or maintenance of, the State II or Stage III water development projects, ... or any subsequent modification thereof ... to the extent that such modification provides for the diversion and transportation of water in the Little Snake River Basin for storage or use in said basin or in Wyoming ... No term or condition shall be imposed ... on the basis of any present or future wilderness" characteristics, designations, or studies in one national forest or in three specified counties.

§ 203—directs management in accordance with the Wilderness Act, subject to valid existing rights.

§ 301(a)—designates three USFS wilderness study areas totaling 180,540 acres in Wyoming. § 301(b)—requires recommendations within three years.

§ 301(c)—directs management of the areas, "subject to valid existing rights and reasonable access to exercise such rights, until Congress determines otherwise ... to maintain their presently existing wilderness character." Withdraws one area from mining laws, and for that area authorizes "oil and gas exploration and development ... under reasonable conditions to protect the environment according to the laws and regulations generally applicable to nonwilderness lands." Allows continued snowmobiling "in the same manner and degree as was occurring prior to the date of enactment."

§ 401—provides statewide USFS release language, except for the area identified in § 5 of P.L. 92- 476.

§ 501—requires a review of grazing "to insure that such policies, practices, and regulations fully conform with and implement the intent of Congress" in the Wilderness Act and P.L. 98-406.

§ 502—establishes no "express or implied claim or denial on the part of the Federal Government as to exemption from Wyoming water laws."

§ 503—directs that state jurisdiction over and responsibilities for fish and wildlife are unaffected. § 504—states that the statute creates no buffer zone.

§ 505—requires, for all wilderness areas in Wyoming, cooperation with the state "in conducting a cultural resources management program ... [for] the protection of archaeological sites and interpretation of such sites for the public benefit and knowledge."

P.L. 98-574—The Texas Wilderness Act of 1984

§ 2—designates five new USFS wilderness areas totaling 34,346 acres in Texas.

§ 4(a)—directs management in accordance with the Wilderness Act, subject to valid existing rights.

§ 4(b)—encourages an expeditious exchange to acquire timber industry land within two of the wilderness areas.

§ 5—provides statewide USFS release language.

P.L. 98-578—The Tennessee Wilderness Act of 1984

§ 2—designates three new USFS wilderness areas totaling 24,942 acres in Tennessee.

§ 4—directs management in accordance with the Wilderness Act, subject to valid existing rights. § 5—provides USFS release language for certain counties in Tennessee. (See P.L. 99-490.)

§ 6(a)—designates two USFS wilderness study areas totaling 5,100 acres in Tennessee, to be reviewed in the initial NFMA Cherokee National Forest management plan.

§ 6(b)—directs management of the wilderness study areas, "until Congress determines otherwise ... so as to maintain its presently existing wilderness character."

P.L. 98-585—The Pennsylvania Wilderness Act of 1984

§ 4—designates two new USFS wilderness areas totaling 9,705 acres in Pennsylvania.

§ 5(a)—directs management in accordance with the Wilderness Act, subject to valid existing rights.

§ 5(b)—directs that state jurisdiction over and responsibilities for fish and wildlife are unaffected. § 8—provides statewide USFS release language.

§ 9—states that the statute creates no buffer zone.

P.L. 98-586—The Virginia Wilderness Act of 1984

§ 2—designates 10 new USFS wilderness areas and one USFS wilderness area addition totaling 55,984 acres in Virginia.

§ 4—directs management in accordance with the Wilderness Act, subject to valid existing rights. § 5—provides statewide USFS release language, except for areas in § 6.

§ 6(a)—designates four USFS wilderness study areas totaling 25,075 acres in Virginia.

§ 6(c)—directs management of the areas, "subject to valid existing rights [and] ... until Congress determines otherwise ... so as to maintain its presently existing wilderness character."

§ 6(d)—requires a study within two years, by the state and the U.S. Environmental Protection Agency, of "the effects of the proposed industrial development site at Covington, Virginia, on air quality off the areas designated for wilderness study."

§ 7—states that the statute creates no buffer zone.

P.L. 98-603—The San Juan Basin Wilderness Protection Act of 1984

§ 102(a)—designates two new BLM wilderness areas totaling 27,840 acres in New Mexico.

§ 102(b)—directs management in accordance with the Wilderness Act, subject to valid existing rights.

§ 102(d)—allows continued livestock grazing "where established before the date of enactment ... subject to such reasonable regulations, policies, and practices as the Secretary deems necessary,

as long as such regulations, policies, and practices fully conform with and implement the intent of Congress regarding grazing in such areas as ... expressed in the Wilderness Act."

§ 109—designates one USFS wilderness area addition with 20 acres in New Mexico; authorizes the "continuance of the existing diversion dam and existing related facilities ... including the provision for access and the use of mechanized equipment only for construction and maintenance of existing structures" and allows "upgrading of the existing diversion dam ... within four years."

P.L. 99-197—The Kentucky Wilderness Act of 1985

§ 2—designates one new USFS wilderness area with 13,300 acres in Kentucky.

§ 3—directs management in accordance with the Wilderness Act, subject to valid existing rights. § 4—provides statewide USFS release language.

P.L. 99-490—The Tennessee Wilderness Act of 1986

§ 2—designates four new USFS wilderness areas and two USFS wilderness area additions totaling 33,735 acres in Tennessee.

§ 4—directs management in accordance with the Wilderness Act, subject to valid existing rights. § 5—provides USFS release language for the remaining counties in Tennessee. (See P.L. 98-578.)

P.L. 99-504—The Nebraska Wilderness Act of 1985

§ 101—designates one new USFS wilderness area with 8,100 acres in Nebraska.

§ 102(a)—directs management in accordance with the Wilderness Act, subject to valid existing rights.

§ 102(b)(1)—allows continued livestock grazing, in accordance with the Wilderness Act and § 108 of P.L. 96-560.

§ 102(b)(2)—requires a review of all grazing management "to insure that such policies, practices, and regulations fully conform with and implement the intent of Congress."

§ 104—provides statewide USFS release language.

P.L. 99-555—The Georgia Wilderness Act of 1986

§ 2—designates four new USFS wilderness areas and one USFS wilderness area addition totaling 42,258 acres in Georgia.

§ 4—directs management in accordance with the Wilderness Act, subject to valid existing rights.

P.L. 99-584—The Texas Wilderness Act Amendments of 1986

§ 2(a)—modifies the boundaries of the areas designated in P.L. 98-574 by reference to a new map.

§ 2(b)—directs management in accordance with the Wilderness Act, subject to valid existing rights; allows continued livestock grazing "subject to such reasonable regulations as are deemed necessary."

P.L. 99-635—Olympic National Park Boundary Adjustment

§ 1(c)—replaces existing designations for three USFS wilderness areas in Washington, deleting 4,571 acres. (This was temporary; Olympic National Park wilderness was designated in P.L. 100- 668.)

P.L. 100-184—The Michigan Wilderness Act of 1987

§ 3—designates 10 new USFS wilderness areas totaling 91,535 acres in Michigan.

§ 5—directs management in accordance with the Wilderness Act, subject to valid existing rights. § 6—provides statewide USFS release language.

§ 7—states that the statute creates no buffer zone.

§ 8—directs that state jurisdiction over and responsibilities for fish and wildlife are unaffected.

§ 9—assures the "right of adequate access" to private lands and requires "concurrence of the owner" for the acquisition of private lands.

§ 10—allows measures to control fire, insects, and diseases "as provided in ... the Wilderness Act."

P.L. 100-225—El Malpais National Monument, Title IV

§ 401—designates two new BLM wilderness areas totaling 98,210 acres in New Mexico.

§ 402(a)—directs management in accordance with the Wilderness Act, subject to valid existing rights.

§ 402(b)—allows continued livestock grazing "subject to reasonable regulations, policies, and practices" that "fully conform with and implement the intent of Congress" in the Wilderness Act and § 108 of P.L. 96-560.

P.L. 100-326—Virginia Wilderness

§ 1—designates four new USFS wilderness areas and two USFS wilderness area additions totaling 27,687 acres in Virginia.

§ 2—directs management in accordance with the Wilderness Act, subject to valid existing rights. § 4—repeals the wilderness study section (§ 6) of P.L. 98-586.

P.L. 100-499—The Winding Stair Mountain National Recreation Area and Wilderness Area Act

§ 3—designates two new USFS wilderness areas totaling 13,954 acres in Oklahoma.

§ 5—directs management in accordance with the Wilderness Act, subject to valid existing rights. § 6—provides statewide USFS release language for Oklahoma.

§ 7—states that the statute creates no buffer zone.

§ 16—allows measures to control fire, insects, and diseases "as provided in ... the Wilderness Act."

§ 18—directs that state jurisdiction over and responsibilities for fish and wildlife are unaffected.

P.L. 100-524—The Congaree Swamp National Monument [Now National Park] Expansion and Wilderness Act

§ 2(a)—designates one new NPS wilderness area with 15,010 acres in South Carolina.

§ 2(b)—designates 6,840 acres of potential NPS wilderness in South Carolina; adds the area to wilderness when nonfederal lands have been acquired and "all uses thereon prohibited by the Wilderness Act have ceased."

§ 4—directs management in accordance with the Wilderness Act, subject to valid existing rights.

P.L. 100-547—The Sipsey Wild and Scenic River and Alabama Addition Act of 1988

§ 201(a)—designates two USFS wilderness area additions totaling 13,970 acres in Alabama.

§ 201(b)—directs management in accordance with the Wilderness Act, subject to valid existing rights.

§ 201(c)—allows measures to control fire, insects "including the Southern Pine Beetle," and diseases "subject to conditions as the Secretary determines desirable," including "such measures in the Secretary's judgment ... deemed necessary to protect threatened resources on Federal, State, or private adjacent lands."

§ 201(d)—allows the Secretary to "convert existing roads ... into suitable hiking or horse trails." § 202—provides statewide USFS release language.

P.L. 100-668—The Washington Park Wilderness Act of 1988

§ 101—designates one new NPS wilderness area with 876,669 acres in Washington.

§ 102—allows the Secretary to "upgrade, maintain and replace, as necessary, the Wolf Creek underground powerline to Hurricane Ridge: *Provided*, That to the extent practicable, such maintenance and operation shall be conducted in such a manner as to remain consistent with wilderness management."

§ 201—designates one new NPS wilderness area with 634,614 acres and potential wilderness additions of 5,226 acres in Washington.

§ 202—allows continued operation of specified hydroelectric projects.

§ 301—designates one new NPS wilderness area with 216,855 acres in Washington.

§ 302—allows the Secretary to "upgrade, maintain and replace as necessary, the Paradise powerline from Longmire to Paradise: *Provided*, That to the extent practicable, such maintenance and operation shall be conducted in such a manner as to protect scenic viewsheds."

Title IV (a)(1)—directs management in accordance with the Wilderness Act, subject to valid existing rights.

Title IV (a)(2)—requires administration of potential wilderness additions "insofar as practicable as wilderness" and adds the areas upon *Federal Register* notice "that all uses thereon that are inconsistent with the Wilderness Act have ceased or that non-Federal interests in land have been acquired."

Title IV (a)(3)—states that the statute creates no buffer zone.

Title V, § 502—"subject to valid existing rights ... Congress hereby expressly reserves such water rights as necessary, for the purposes for which such areas are so designated."

P.L. 101-195—The Nevada Wilderness Protection Act of 1989

§ 2—designates 13 new USFS wilderness areas and one USFS wilderness area addition totaling 733,400 acres in Nevada.

§ 4—directs management in accordance with the Wilderness Act, subject to valid existing rights. § 5—provides statewide USFS release language.

§ 6(a)—allows continued livestock grazing "where established before the date of enactment" in accordance with the Wilderness Act and P.L. 96-560.

§ 6(b)—requires a review of all grazing management "to insure that such policies, practices, and regulations fully conform with and implement the intent of Congress."

§ 7—states that the statute creates no buffer zone.

§ 8—"reserves a quantity of water sufficient to fulfill the purposes." Requires the Secretary to act to protect those rights.

§ 9—directs that state jurisdiction over and responsibilities for fish and wildlife are unaffected.

§ 10—authorizes, "where appropriate, the installation and maintenance of hydrologic, meteorologic, or climatological collection devices ... where such facilities and access thereto are essential to flood warning, flood control and water reservoir operation," subject to reasonable terms and conditions.

§ 11—states that the statute is to have no effect on low-level military overflights, new units of special use airspace, and flight training routes over four specific wilderness areas.

P.L. 101-401—The Maine Wilderness Act of 1990

§ 2—designates one new USFS wilderness area with 12,000 acres in Maine.

§ 4—directs management in accordance with the Wilderness Act, subject to valid existing rights. § 5—provides statewide USFS release language.

§ 6—states that the statute creates no buffer zone.

§ 7—allows measures to control fire, insects, and diseases "as provided in ... the Wilderness Act."

§ 8—directs that state jurisdiction over and responsibilities for fish and wildlife are unaffected.

P.L. 101-626—The Tongass Timber Reform Act

§ 202—amends a previous wilderness statute (P.L. 96-487) to designate five new USFS wilderness areas and one USFS wilderness area addition totaling 296,080 acres in Alaska.

P.L. 101-628—The Arizona Desert Wilderness Act of 1990

§ 101(a)—designates 38 new BLM wilderness areas and one BLM wilderness area addition totaling 1,089,970 acres in Arizona.

§ 101(a)(3)—provides that "the existing water pipeline for the town of Oatman, together with the right of ingress and egress thereto, may be operated, maintained, and upgraded, subject to reasonable requirements to protect wilderness values."

§ 101(a)(20)—provides that a specified right-of-way and "the right of ingress or egress thereto shall not be affected... and the existing powerline utilizing such right-of-way may be operated, maintained, and upgraded, subject to reasonable requirements to protect wilderness values."

§ 101(b)—directs management in accordance with the Wilderness Act, subject to valid existing rights.

§ 101(d)—states that the statute creates no buffer zone.

§ 101(e)—directs that state jurisdiction over and responsibilities for fish and wildlife are unaffected.

§ 101(f)(1)—allows continued livestock grazing "where established before the date of enactment" in accordance with the Wilderness Act and Appendix A of H.Rept. 101-405.

§ 101(f)(2)—requires a review of all grazing management "to insure that such policies, practices, and regulations fully conform with and implement the intent of Congress."

§ 101(g)—"reserves a quantity of water sufficient to fulfill the purposes of this title." Requires the Secretary to act to protect those rights.

§ 101(h)—states that the statute is to have no effect on low-level military overflights, new units of special use airspace, and flight training routes.

§ 101(i)—requires mineral rights in the wilderness areas to be acquired by exchange.

§ 101(k)—modifies a previously-enacted provision for a specific wilderness area concerning management and use of a particular road.

§ 101(l)—states that the statute is to have no effect on "the operation for flood control purposes" of a particular dam.

§ 102—provides statewide BLM release language, except for two specified areas.

§ 301(a)—designates four new FWS wilderness areas totaling 1,343,444 acres in Arizona.

§ 301(b)—directs management in accordance with the Wilderness Act, subject to valid existing rights.

§ 301(c)(1)(A)[24]—"reserves a quantity of water sufficient to fulfill the purposes of this title."

§ 301(c)(1)(B)[24]—establishes "no right to water of the Colorado River are reserved, either expressly, impliedly, or otherwise."

§ 301(c)(2)[24]—requires the Secretary to act to protect those rights.

§ 301(c)(3)[24]—directs no "relinquishment or reduction of any water right reserved or appropriated by the United States ... before the date of enactment."

§ 301(e)—states that the statute is to have no effect on "the operation of federally owned dams located on the Colorado River."

§ 301(f)—states that the statute is to have no effect on low-level military overflights or ground instrumentation for the Cabeza Prieta National Wildlife Refuge.

§ 301(g)—states that the statute is to have no effect on "border operations by the Immigration and Naturalization Service, the Drug Enforcement Agency, or the United States Customs Service."

§ 302—directs that nothing in Titles I, II, or III "shall amend, construe, supersede, or preempt any State law, Federal law, interstate compact, or international treaty pertaining to the Colorado River (including its tributaries) in the Upper Basin, including, but not limited to, the appropriation, use, development, storage, regulation, allocation, conservation, exportation, or quality of those waters."

H. Rept. 101-405 to Accompany H.R. 2570 (P.L. 101-626)—Appendix A. Grazing Guidelines

Pages 41-43 of the report supplement the "very clear intent" of § 4(d)(4)(2) of the Wilderness Act and of H.Rept. 96-617 "that livestock grazing, and activities and the necessary facilities to support a livestock grazing program, will be permitted to continue ... when grazing was established prior to classification of an area as wilderness." The guidelines provide direction on five points:

1) Prohibits curtailing grazing simply because an area has been designated as wilderness, and directs that wilderness designation should not be used "as an excuse by administrators to slowly 'phase out' grazing."
2) Allows maintenance of existing supporting facilities (e.g., fences, line cabins, water wells and lines, stock tanks, etc.). Allows occasional use of motorized equipment where practical alternatives do not exist, "based on a rule of practical necessity and reasonableness."
3) Does not require replacement or reconstruction of deteriorated facilities or improvements to use natural materials, "unless the material and labor costs of using natural materials are such that their use would not impose unreasonable additional costs."
4) Allows construction of new improvements or replacement of deteriorated facilities, if in accordance with these guidelines and area management plans and "primarily for the purpose of resource protection and the more effective management of these resources."
5) Allows motorized equipment for emergencies "such as rescuing sick animals or the placement of feed in emergency situations."

H. Rept. 101-405 to Accompany H.R. 2570 (P.L. 101-626)—Appendix B. Wildlife Management Guidelines

Pages 44-51 present the committee's conclusions of applicable, previously existing BLM policies and guidelines for fish and wildlife management in wilderness areas. After purposes and general policy, the report includes 14 points authorizing various activities, with multiple guidelines for implementing several of the points:.

1) Allows motorized equipment if "truly necessary to administer." Expects "that any such use should be rare and temporary; that no

roads can be built; and that ... such use is the minimum necessary to accomplish the task."

2) Allows fish and wildlife research and management surveys when conducted "in a manner compatible with the preservation of the wilderness environment." Methods that "temporarily infringe on the wilderness environment may be approved if alternative methods or other locations are not available." Aircraft use must "minimizes disturbance of other users, including humans and wildlife."
3) Allows facility development and habitat alteration "in rare instances ... to alleviate adverse impacts caused by human activities." Allows existing flow-maintenance dams, water developments, water diversions, ditches, and other fish and wildlife habitat developments. Allows "maintenance of existing water supplies and development of additional water supplies ... but only when essential to preserve the wilderness resource and to correct unnatural conditions resulting from human influence."
4) Allows actions necessary to protect or recover threatened or endangered species, "including habitat manipulation and special protection measures." Directs using "only the minimum actions necessary and the methods most appropriate in wilderness."
5) Allows angling, hunting and trapping, "subject to applicable State and Federal laws and regulations."
6) Allows scientific sampling of fish and wildlife populations in wilderness.
7) Allows chemical treatment "to prepare waters for reestablishment of indigenous species, to protect or recover Federally listed threatened or endangered species, or to correct undesirable conditions resulting from the influence of man. Species of fish traditionally stocked before wilderness designation may be considered indigenous."
8) Allows collecting fish spawn "when alternative sources are unavailable or unreliable, or where spawn taking was an established practice before wilderness designation."
9) Allows fish stocking "using means appropriate for wilderness," to reestablish or maintain an indigenous species, or (b) to perpetuate or recover a threatened or endangered species. Prohibits stocking with exotic species.[25] Allows stocking of barren lakes or streams, "if there is mutual agreement that no appreciable loss of scientific values or adverse effects on wilderness resources will occur."

10) Allows aerial fish stocking "where this was an established practice before the wilderness designation or where other practical means are not available."
11) Allows wildlife transplants (removal, reintroduction, or supplemental introduction) "if necessary: (a) to perpetuate or recover a threatened or endangered species; of (b) to restore the population of an indigenous species eliminated or reduced by human influence." Directs transplants "in a manner compatible with the wilderness character of the area."
12) Allows "wildlife damage control," including for non-indigenous species, "to protect Federally listed threatened or endangered species, to prevent transmission of diseases or parasites affecting other wildlife and humans, or to prevent serious losses of domestic livestock."
13) Allows control of visitor access and use to protect wildlife "sensitive to human encroachments on their ranges ... particularly during certain seasons of the year."
14) Directs fire management to "(a) permit lightning-caused fires to play, as nearly as possible, their natural ecological role within wilderness and (b) reduce, to an acceptable level, the risks and consequences of wildfire within wilderness or escaping from wilderness."

P.L. 101-633—The Illinois Wilderness Act of 1990

§ 3—designates seven new USFS wilderness areas and one USFS wilderness area addition totaling 26,266 acres in Illinois.

§ 5—directs management in accordance with the Wilderness Act, subject to valid existing rights. § 6—states that the statute creates no buffer zone.

§ 7—directs that state jurisdiction over and responsibilities for fish and wildlife are unaffected. § 8—allows measures to control fire, insects, and diseases.

§ 9—permits relatives and descendents, and those accompanying them, access to and maintenance of cemeteries, limited to "appropriate access and maintenance to minimize any detrimental effects on the wilderness resource."

P.L. 102-217—The Chattahoochee National Forest Protection Act of 1991

§ 2(a)—designates two new USFS wilderness areas and one USFS wilderness area addition totaling 25,840 acres in Georgia.

§ 2(b)—directs management in accordance with the Wilderness Act, subject to valid existing rights.

P.L. 102-301—The Los Padres Condor Range and River Protection Act

§ 2—designates four new USFS wilderness areas and three USFS wilderness area additions totaling 400,450 acres, and one potential USFS wilderness addition with 50 acres in California.

§ 2(5)—directs that "the Toad Springs road corridor delineated as potential wilderness shall remain open to off road vehicle traffic until construction of an alternative route which bypasses this area is completed. These potential wilderness lands shall be automatically incorporated in ... [the] Wilderness upon publication of a notice in the Federal Register."

§ 3(a)—directs management in accordance with the Wilderness Act, subject to valid existing rights.

§ 3(b)—allows that "to guarantee the continued viability of the watersheds ... and to ensure the continued health and safety of the communities serviced by such watershed, the Secretary of Agriculture may take such measures as are necessary for fire prevention and watershed protection including, but not limited to, acceptable fire presuppression and fire suppression measures and techniques."

§ 3(c)—allows "management activities to maintain or restore fish and wildlife populations, including the California condor, and the habitats ... where consistent with relevant wilderness management plans in accordance with appropriate policies and guidelines such as those set forth in Policies and Guidelines for Fish and Wildlife Management in National Forests and Bureau of Land Management Wilderness, dated August 25, 1986."

§ 3(d)—states that the statute creates no buffer zone.

§ 3(e)—"reserves a quantity of water sufficient to fulfill the purposes of this Act." Directs the Secretary to act to protect those rights.

§ 5—provides USFS release language for specified areas.

P.L. 103-77—The Colorado Wilderness Act of 1993

§ 2(a)—designates nine new USFS wilderness areas and 10 USFS wilderness area additions totaling 611,730 acres in Colorado.[26]

§ 2(a)(8)—authorizes acquisition of mineral reservations "only by donation or exchange."

§ 2(a)(13)—establishes "no right, or claim of right, to the diversion and use of the waters by the Fryingpan-Arkansas Project shall be prejudiced, expanded, diminished, altered, or affected by this Act, nor shall anything in this Act be construed to expand, abate, impair, impede, limit, interfere with, or prevent the construction, operation, use, maintenance, or repair of the project facilities and diversion systems to their full extent."

§ 3(a)—directs management in accordance with the Wilderness Act, subject to valid existing rights.

§ 3(b)—allows continued livestock grazing in accordance with the Wilderness Act, § 108 of P.L. 96-560, and Appendix A of H.Rept. 101-405.

§ 3(c)—directs that state jurisdiction over fish and wildlife is unaffected. § 3(e)—states that the statute creates no buffer zone.

§ 3(g)(1)—states that the statute is to have no effect on valid existing rights. § 4—repeals the wilderness study provisions of P.L. 96-560.

§ 8(b)(1)—prohibits any federal employee from asserting, and any court or agency from considering, "any claim to or for water or water rights" based on wilderness designation "as constituting an express or implied reservation of water or water rights."

§ 8(c)—directs that "neither the President nor any" other federal employee "shall fund, assist, authorize, or issue a license or permit for the development of any new water resource facility" as defined in § 8(a)(3) within the wilderness areas.

§ 8(d)(1)—allows "reasonable access to water resource facilities in existence on the date of enactment ... including motorized access where necessary and customarily employed on routes existing on the date of enactment."

§ 8(d)(2)—allows existing, customarily used access routes to be "used, maintained, repaired, and replaced to the extent necessary to maintain their present function, design, and serviceable operation, so long as such activities have no increased adverse impacts on the resources and values."

§ 8(d)(3)—allows existing water resource facilities "to be used, operated, maintained, repaired, and replaced to the extent necessary for the continued exercise ... of vested water rights."

§ 8(d)(4)—allows water resource facilities and access routes to be "maintained and repaired when and to the extent necessary to prevent increased adverse impacts on the resources and values."

§ 8(e)—states that the statute is to have no effect or limitation on "the use, operation, maintenance, repair, modification, or replacement of water resource facilities in existence on the date of enactment."

§ 8(f)—requires monitoring of "the operation of and access to water resource facilities."

§ 8(g)(1)—directs that nothing "shall be construed as limiting, altering, modifying, or amending any of the interstate compacts or equitable apportionment decrees."

§ 8(g)(1)—prohibits any federal employee from asserting, and any court or agency from considering, any claim to or "for waters of the North Platte River" based on the designation of a particular wilderness area .

P.L. 103-433—The California Desert Protection Act of 1994

Title I—Designation of Wilderness Areas to be Administered by the Bureau of Land Management

§ 102—designates 66 new BLM wilderness areas and three BLM wilderness area additions totaling 3,667,020 acres in California.

§ 102(1)—authorizes a right-of-way and road construction solely for installation of a space energy laser facility, if requested by the Secretary of the Navy within 15 years of enactment. "So far as practicable, any such road shall be aligned in a manner that takes into account the desirability of minimizing adverse impacts on wilderness values."

§ 102(13)—authorizes a right-of-way and road construction within the identified "nonwilderness road corridor" upon application from the Department of Defense.

§ 103(a)—directs management in accordance with the Wilderness Act, subject to valid existing rights.

§ 103(c)—allows continued livestock grazing "if established before the date of enactment ... subject to such reasonable regulations, policies, and practices as the Secretary deems necessary," in accordance with the Wilderness Act and § 101(f) of P.L. 101-628.

§ 103(d)—states that the statute creates no buffer zone.

§ 103(e)—directs that state jurisdiction over fish and wildlife is unaffected.

§ 103(f)—authorizes "management activities to maintain or restore fish and wildlife populations and the habitats," including "the use of motorized vehicles by the appropriate State agencies."

§ 103(g)—states that the statute is to have no effect on "Federal, State, and local law enforcement agencies from conducting law enforcement and border operations as permitted before the date of enactment ... including the use of motorized vehicles and aircraft."

§ 104(a)—provides BLM release language for specified areas; (b) retains wilderness-like management "that does not impair the suitability of such areas for preservation as wilderness" for seven BLM wilderness study areas totaling 315,230 acres in California.

§ 104(c)—withdraws the wilderness study areas from public land laws, mining laws, mineral and geothermal leasing laws, and mineral material disposition, subject to valid existing rights.

§ 105—designates one BLM wilderness study area with 11,200 acres in California.

§ 106—requires a report on "current and planned exploration, development or mining activities" in four additional proposed wilderness areas within 10 years of enactment.

Title II—Designation of Wilderness Areas to be Administered by the United States Fish and Wildlife Service

§ 201(a)—designates three new FWS wilderness areas totaling 9,031 acres in California.

§ 201(b)—directs management in accordance with the Wilderness Act, subject to valid existing rights.

§ 202—states that the statute is to have no effect on "the operation of federally owned dams located on the Colorado River."

§ 203—provides that nothing "shall amend, construe, supersede, or preempt any State law, Federal law, interstate compact, or international treaty pertaining to the Colorado River (including its tributaries) in the Upper Basin, including, but not limited to, the appropriation, use, development, storage, regulation, allocation, conservation, exportation, or quality of those waters."

§ 204—provides that this title establishes "no rights to water of the Colorado River are reserved, either expressly, impliedly, or otherwise."

Title VI—National Park Wilderness

§ 601(a)—designates three new NPS wilderness areas totaling 3,985,018 acres in California.

§ 601(b)—designates one potential NPS wilderness with 6,840 acres in California to be added "upon cessation of all uses prohibited by the Wilderness Act and publication in the Federal Register of a notice;" directs management "insofar as practicable as wilderness."

§ 603—directs management in accordance with the Wilderness Act.

Title VII—Miscellaneous Provisions

§ 706—"reserves a quantity of water sufficient to fulfill the purposes of this Act." Directs the Secretary to act to protect those rights.

Title VIII—Military Lands and Overflights

§ 802(a) & (b)—states that there is no effect from "this Act, the Wilderness Act, or other land management laws generally applicable to" the NPS and wilderness areas on "low-level overflights of military aircraft" or "designation of new units of special airspace or the use of establishment of military flight training routes."

P.L. 104-208 (Omnibus Consolidated Appropriations Act, FY1997), Division B—the Oregon Resource Conservation Act of 1996

Title I—The Opal Creek Wilderness and Scenic Recreation Area Act of 1996

§ 104(a)—designates one new USFS wilderness area and transfers part of an existing wilderness to the new area in Oregon.

§ 104(b)—establishes conditions for the designation and transfer.

Title VII—Oregon Islands Wilderness, Additions

§ 701(a)—designates one FWS wilderness area addition with 95 acres in Oregon.

§ 701(c)—designates "all federally owned named, unnamed, surveyed and unsurveyed rocks, reefs, islets and islands lying within three geographic miles off the coast of Oregon and above mean high tide" as part of the wilderness, and transfers jurisdiction to the FWS.

P.L. 104-333—The Omnibus Parks and Public Lands Management Act of 1996

§ 302—Anaktuvuk Pass Land Exchange

(c)(1)—modifies the designation of one NPS wilderness area in Alaska, deleting 73,993 acres and adding 56,825 acres (a net deletion of 17,168 acres).

§ 1022—The Bisti/De-Na-Zin Wilderness Expansion and Fossil Forest Protection Act

(b)(1)—designates one BLM wilderness area addition with 16,525 acres in New Mexico.

(b)(4)—amends P.L. 98-603 by adding a new § 102(e) withdrawing the areas from the mining laws, mineral leasing laws, geothermal leasing laws, and mineral material sales laws, but allowing the exchange of coal leases and oil and gas leases, and authorizing for existing oil and gas leases "terms, stipulations, and conditions ... necessary to avoid significant disturbance of the land surface or impairment of the ecological, educational, scientific, recreational, scenic, and other wilderness values of the lands."

§ 1023—The Opal Creek Wilderness and Scenic Recreation Area

Reenacts Title I of Division B of P.L. 104-208.

§ 1027—Oregon Islands Wilderness, Additions

Reenacts Title VII of Division B of P.L. 104-208.

P.L. 105-75—Eagles Nest Wilderness Boundary Adjustment

§ 1(a)—designates one USFS wilderness area addition with 160 acres in Colorado, pending the acquisition of the parcel before December 31, 2000.

P.L. 105-277 (Omnibus Consolidated and Emergency Supplemental Appropriations Act, 1999), Division A, § 101(e) (Interior and Related Agencies)

§ 610—designates one USFS wilderness study area of 15,000 acres in Washington, pending the consummation of a land exchange, with recommendations to the President within three years, management of the lands, "subject to valid existing rights and existing uses ... until Congress determines otherwise or until December 31, 2003, ... to maintain their wilderness character," and withdraws the lands from mining and mineral leasing (including geothermal leasing) laws.

P.L. 106-65 (FY2002 National Defense Authorization Act)

§ 3032(b)(1)—directs management the previously-established FWS Cabeza Prieta Wilderness "(A) for the purposes for which the refuge and wilderness were established; and (B) to support current and future military aviation training needs."

§ 3032(c) & (d)—extend and amend an existing memorandum of understanding to continue and expand military aviation and ground instrumentation with few limitations.

§ 3032(e)—authorizes public closures of the refuge and wilderness for public safety and national security.

P.L. 106-76—The Black Canyon of the Gunnison National Park and Gunnison Gorge National Conservation Area Act of 1999

§ 6(a)—adds 4,419 acres to an existing NPS wilderness in Colorado.

§ 8(a)(1)—designates one new BLM wilderness area with 17,700 acres in Colorado.

§ 8(b)—directs management in accordance with the Wilderness Act, subject to valid existing rights.

§ 8(c)—directs that state jurisdiction over fish and wildlife is unaffected.

§ 10(a)—establishes no "express or implied reservation of water for any purpose" or effect on existing water rights.

§ 10(b)—requires establishing "any new water right that ... is necessary for the purposes of this Act ... in accordance with the procedural and substantive requirements" of the state.

P.L. 106-145—The Otay Mountain Wilderness Act of 1999

§ 3—designates one new BLM wilderness area with 18,500 acres in California. § 5—provides BLM release language for specified areas.

§ 6(a)—directs management in accordance with the Wilderness Act, subject to valid existing rights.

§ 6(b)—authorizes "drug interdiction, border operations, and wildland fire management operations," in accordance with the Wilderness Act and "subject to such conditions as the Secretary considers appropriate."

§ 8—states that the statute creates no buffer zone.

P.L. 106-156—The Dugger Mountain Wilderness Act of 1999

§ 2(a)—designates one new USFS wilderness area with 9,200 acres in Alabama.

§ 2(c)—directs management in accordance with the Wilderness Act, subject to valid existing rights.

§ 2(d)—allows "ground-based mechanical and motorized equipment to disassemble and remove" the fire tower and supporting structures. The road to the fire tower is to be closed after the tower is removed.

P.L. 106-353—The Colorado Canyons National Conservation Area and Black Ridge Canyons Wilderness Act of 2000

§ 5—designates one new BLM wilderness area with 75,550 acres in Colorado.

§ 6(c)—withdraws the areas from public land laws, mining laws, mineral and geothermal leasing laws, and mineral material disposition, subject to valid existing rights.

§ 6(e)—directs management in accordance with the Wilderness Act, subject to valid existing rights.

§ 6(f)—authorizes hunting, trapping, and fishing; allows zone and period closures for "public safety, administration, or public use and enjoyment."

§ 6(g)(2)—allows livestock grazing, consistent with the Wilderness Act and Appendix A of H.Rept. 101-405.

§ 6(h)—requires "a comprehensive management plan for the long-range protection and management of the Conservation Area and the Wilderness" within three years.

§ 6(i)—states that the statute creates no buffer zone.

§ 6(l)(2)—directs that nothing shall (A) constitute "either an express or implied reservation of any water or water rights;" (B) affect any existing "conditional or absolute water rights in the State;" (C) be construed as "limiting, altering, modifying, or amending any of the interstate compacts or equitable apportionment decrees that apportion water."

§ 9—directs the Secretary (a) to "continue to allow private landowners reasonable access to inholdings;" and (b) to grant "public right of access, including commercial vehicles, to Glade Park."

P.L. 106-399—The Steens Mountain Cooperative Management and Protection Act of 2000

§ 201—designates one new BLM wilderness area with 170,000 acres in Oregon. § 202(a)—directs management in accordance with the Wilderness Act.

§ 202(c)—requires "reasonable access to private lands within the boundaries."

§ 202(d)(1)—allows continued livestock grazing, consistent with the Wilderness Act and Appendices A and B of H.Rept. 101-405.

§ 202(d)(2)—directs permanent retirement of grazing permits for identified lands.

§ 203—directs that "nothing shall constitute an express or implied claim or denial on the part of the Federal Government as to exemption from State water laws."

§ 204—states that the statute is to have no effect on existing wilderness study areas; (c) expands one existing BLM wilderness study area.

P.L. 106-456—The Spanish Peaks Wilderness Act of 2000

§ 2—amends P.L. 103-77 to designate one new USFS wilderness area with 18,000 acres in Colorado.

§ 3(a)—directs continued "historic uses of the Bulls Eye Mine Road ... subject to such terms and conditions as the Secretary may provide."

§ 3(b)—requires access to privately owned land within the wilderness in accordance with the Wilderness Act.

P.L. 106-471—Virginia Wilderness

§ 1—amends P.L. 100-326 to designate two new USFS wilderness areas totaling 10,571 acres in Virginia.

P.L. 106-554 (Consolidated Appropriations Act, 2004), incorporating by reference the Black Rock Desert-High Rock Canyon Emigrant Trails National Conservation Area Act of 2000

§ 8(a)—designates 10 new BLM wilderness areas totaling 757,500 acres in Nevada.

§ 8(b)—directs management in accordance with the Wilderness Act, subject to valid existing rights.

§ 8(d)—allows continued livestock grazing, subject to "reasonable regulations, policies, and practices" that conform with the Wilderness Act, § 101(f) of P.L. 101-628.

P.L. 107-107 (FY2002 National Defense Authorization Act)

§ 1049—authorizes closure of the Vieques Naval Training Range, Puerto Rico; transfers the area to the Department of the Interior as part of the National Wildlife Refuge System; designates the 900-acre "Live Impact Area" as wilderness; and closes the wilderness to access by the public.

P.L. 107-206 (Supplemental Appropriations Act for Further Recovery From and Response to Terrorist Attacks on the United States)

§ 706(n)—amends § 103 of P.L. 96-560 to designate one USFS wilderness area addition with 3,600 acres in South Dakota.

P.L. 107-216—The James Peak Wilderness and Protection Area Act

§ 2(a)—amends P.L. 103-77 to designate one new USFS wilderness area and two USFS wilderness area additions totaling 17,195 acres in Colorado.

P.L. 107-282—The Clark County Conservation of Public Land and Natural Resources Act of 2002

§ 202(a)—designates 17 new BLM wilderness areas and one BLM wilderness area addition totaling 451,915 acres in Nevada.

§ 202(d)—withdraws the areas from public land laws, mining laws, mineral leasing laws, mineral materials laws, and geothermal leasing laws.

§ 203(a)—directs management in accordance with the Wilderness Act, subject to valid existing rights.

§ 203(b)—allows continued livestock grazing, subject to "reasonable regulations, policies, and practices" that conform with the Wilderness Act and Appendix A of H.Rept. 101-405.

§ 203(d)(2)—states that nothing in the act "(A) ... shall constitute or be construed to constitute either an express or implied reservation by the United States of any water or water rights ... (B) ... shall affect any water rights in the State ... (C) ... shall be construed as establishing a precedent ... (D) ... shall be construed as limiting, altering, modifying, or amending any interstate compacts or equitable apportionment decrees that apportion water among and between the State of Nevada and other States; [and] (E) ... construed as limiting, altering, modifying, or amending the Clark County Multi-Species Habitat Conservation Plan."

§ 203(d)(3)—directs the Secretary to "follow the procedural and substantive requirements of the law of the State ... to obtain and hold any water rights."

§ 203(d) (4)(B)—prohibits funding, assisting, authorizing, or permitting for new water resource facilities within the wilderness areas.

§ 204—states that the statute creates no buffer zone.

§ 205(a)—states that the statute is to have no effect on low-level military overflights, flight testing or evaluation, or new units of special use airspace.

§ 206—prohibits diminution of tribal access rights for tribal activities, "including spiritual, cultural, and traditional food-gathering activities."

§ 207—provides BLM release language for specified areas.

§ 208(a)—directs that state jurisdiction over fish and wildlife is unaffected.

§ 208(b)—authorizes "management activities to maintain or restore fish and wildlife populations and the habitats to support such populations ... in accordance with appropriate policies" from Appendix B of H.Rept. 101-405, "including the occasional and temporary use of motorized vehicles if such use ... would promote healthy, viable, and more naturally distributed wildlife populations that would enhance wilderness values ... with the minimum impact necessary."

§ 208(c)—authorizes continued use of aircraft "to survey, capture, transplant, monitor, and provide water for wildlife populations, including bighorn sheep, and feral stock, horses, and burros" consistent with the Wilderness Act and Appendix B of H.Rept. 101-405.

§ 208(d)—authorizes "structures and facilities ... for wildlife water development projects, including guzzlers" if they enhance the wilderness values and "the visual impacts ... can reasonably be minimized."

§ 208(e)—authorizes, in consultation with the state, zones and periods of no hunting, fishing, or trapping "for public safety, administration, or compliance with applicable laws."

§ 208(f)—directs a cooperative agreement with the state specifying "the terms and conditions under which the State ... may use wildlife management activities in the wilderness areas."

§ 209—directs that nothing "precludes ... wildfire management operations (including operations using aircraft or mechanized equipment) to manage wildfires in the wilderness areas."

§ 210—directs that nothing "precludes ... the installation and maintenance of hydrologic, meteorologic, or climatological collection devices ... if the facilities and access ... are essential to flood warning, flood control, and water reservoir operation," subject to prescribed terms and conditions.

§ 211—states that the statute is to have no effect on the "laws, regulations, or management policies applicable to the National Park Service for Lake Mead National Recreation Area."

P.L. 107-370—The Big Sur Wilderness and Conservation Act of 2002

§ 2(a)—designates 13 USFS and BLM wilderness area additions totaling 56,880 acres in California.

§ 3(a)—directs management in accordance with the Wilderness Act, subject to valid existing rights.

§ 3(b)—authorizes livestock grazing, consistent with the Wilderness Act, § 108 of P.L. 96-560, and Appendix A of H.Rept. 101-405.

§ 3(c)—directs that state jurisdiction over fish and wildlife is unaffected.

§ 3(d)—"reserves a quantity of water sufficient to fulfill the purposes of this Act." Directs the Secretary to act to protect those rights.

§ 4—directs management plan amendments to authorize "whatever appropriate actions ... are necessary for fire prevention and watershed protection consistent with wilderness values."

§ 5(a)—states that the statute is to have no effect on low-level military overflights, new units of special use airspace, or military flight training routes.

§ 5(b)—authorizes continued "nonmotorized access to and use of the wilderness areas ... for military training ... in the same manner and degree as authorized prior to enactment."

§ 7—authorizes "the construction and maintenance of a new water line and corresponding spring box improvements adjacent to an existing domestic water service" in one wilderness area.

P.L. 108-424—The Lincoln County Conservation, Recreation, and Development Act of 2004, Title II

§ 203(a)—designates 14 new BLM wilderness areas totaling 768,294 acres in Nevada.

§ 203(d)—withdraws the areas from public land laws, mining laws, mineral leasing laws, and geothermal leasing laws.

§ 204(a)—directs management in accordance with the Wilderness Act, subject to valid existing rights.

§ 204(b)—allows continued livestock grazing, subject to "reasonable regulations, policies, and practices" that are consistent with the Wilderness Act and Appendix A of H.Rept. 101-405.

§ 204(d)(2)—directs that nothing "(A) shall constitute or be construed to constitute either an express or implied reservation by the United States of any water or water rights ... (B) shall affect any water rights in the State ... (C) shall be construed as establishing a precedent ... or (E) shall be construed as limiting, altering, modifying, or amending any of the interstate compacts or equitable apportionment decrees that apportion water among and between the State and other States."

§ 204(d)(3)—directs the Secretary to “follow the procedural and substantive requirements of the law of the State ... to obtain and hold any water rights.”

§ 204(d)(4)(B)—prohibits funding, assisting, authorizing, or permitting for new water resource facilities within the wilderness areas.

§ 205—states that the statute creates no buffer zone.

§ 206—states that the statute is to have no effect on low-level military overflights, flight testing or evaluation, or new units of special use airspace.

§ 207—prohibits diminution of tribal access rights for tribal activities, “including spiritual, cultural, and traditional food-gathering activities.”

§ 208—provides BLM release language for specified areas.

§ 209(a)—directs that state jurisdiction over fish and wildlife is unaffected.

§ 209(b)—authorizes “management activities to maintain or restore fish and wildlife populations and the habitats to support such populations ... in accordance with appropriate policies” from Appendix B of H.Rept. 101-405, “including the occasional and temporary use of motorized vehicles if such use ... would promote healthy, viable, and more naturally distributed wildlife populations that would enhance wilderness values ... with the minimum impact necessary.”

§ 209(c)—authorizes continued use of aircraft “to survey, capture, transplant, monitor, and provide water for wildlife populations, including bighorn sheep, and feral stock, horses, and burros” consistent with the Wilderness Act and Appendix B of H.Rept. 101-405.

§ 209(d)—authorizes “structures and facilities ... for wildlife water development projects, including guzzlers” if they enhance the wilderness values and “the visual impacts ... can reasonably be minimized.”

§ 209(e)—authorizes, in consultation with the state, zones and periods of no hunting, fishing, or trapping “for public safety, administration, or compliance with applicable laws.”

§ 209(f)—requires a cooperative agreement with the state specifying “the terms and conditions under which the State ... may use wildlife management activities in the wilderness areas.”

§ 210—directs that nothing “precludes ... wildfire management operations (including operations using aircraft or mechanized equipment) to manage wildfires in the wilderness areas.”

§ 211—directs that nothing “precludes ... the installation and maintenance of hydrologic, meteorologic, or climatological collection devices ... if the facilities and access ... are essential to flood warning, flood control, and water reservoir operation" subject to prescribed terms and conditions.

P.L. 108-447 (Consolidated Appropriations Act, FY2005), Division E (Interior and Related Agencies)

§ 140(c)—designates one new NPS wilderness area with 33,500 acres in Wisconsin.

§ 140(d)(1)—directs management in accordance with the Wilderness Act, subject to valid existing rights.

§ 140(d)(2)—states that the statute is to have no effect on "any treaty rights;" management of the waters of Lake Superior; or "the use of motors on the lake waters, including snowmobiles and the beaching of motorboats ... and the maintenance and expansion of any docks existing at the time of the enactment."

§ 145(a)—replaces § 2 of P.L. 97-250, essentially providing one NPS wilderness area addition of 1,026 acres and reducing the NPS potential wilderness by 1,218 acres in Georgia.

P.L. 109-94—The Ojito Wilderness Act

§ 3(a)—designates one new BLM wilderness area with 11,183 acres in New Mexico.

§ 3(c)—directs management in accordance with the Wilderness Act, subject to valid existing rights.

§ 3(f)—provides BLM release language for specified areas.

§ 3(g)—allows continued livestock grazing "where established before the date of enactment," consistent with the Wilderness Act and Appendix A of H.Rept. 101-405.

§ 3(h)—directs that state jurisdiction over or responsibilities for fish and wildlife are unaffected.

§ 3(i)—directs that nothing "(A) shall constitute or be construed to constitute either an express or implied reservation by the United States of any water or water rights ... (B) shall affect any water rights in the State ... (C) shall be construed as establishing a precedent ... or (E) shall be construed as limiting, altering, modifying, or amending any of the interstate compacts or equitable apportionment decrees that apportion water among and between the State and other States."

§ 3(i)(3)—directs the Secretary to "follow the procedural and substantive requirements of the law of the State ... to obtain and hold any water rights."

§ 3(i)(4)(B)—prohibits funding, assisting, authorizing, or permitting for new water resource facilities within the wilderness areas, except wildlife guzzlers.

§ 3(j)—withdraws areas from public land laws, mining laws, mineral and geothermal leasing laws, and mineral material laws, subject to valid existing rights.

P.L. 109-118—The Caribbean National Forest Act of 2005

§ 3(a)—designates one new USFS wilderness area with 10,000 acres in Puerto Rico.

§ 3(c)—directs management in accordance with the Wilderness Act, subject to valid existing rights.

§ 3(d)—directs that nothing "precludes the installation and maintenance of hydrologic, meteorological, climatological, or atmospheric data collection and remote transmission facilities, or any combination of those facilities ... [if] essential to the scientific research purposes of the Luquillo Experimental Forest."

P.L. 109-163 (National Defense Authorization Act for FY2006), Subtitle H—Utah Test and Training Range

§ 384(a)—designates one new BLM wilderness area with 101,400 acres in Utah.

§ 384(b)—withdraws areas from public land laws, mining laws, mineral and geothermal leasing laws, and mineral material disposition, subject to valid existing rights.

§ 384(d)—directs management in accordance with the Wilderness Act, subject to valid existing rights.

§ 384(f)—directs that state jurisdiction over fish and wildlife is unaffected.

§ 384(g)—allows continued livestock grazing "subject to such reasonable regulations, policies, and practices" that conform with the Wilderness Act, § 101(f) of P.L. 101-628, and Appendix A of H.Rept. 101-405.

§ 384(h)—states that the statute creates no buffer zone.

§ 384(i)—provides BLM release language for specified areas.

P.L. 109-362—The Northern California Coastal Wild Heritage Wilderness Act

§ 3—designates eight new USFS and BLM wilderness areas and four USFS and BLM wilderness area additions totaling 264,559 acres in California.

§ 4(a)—directs management in accordance with the Wilderness Act, subject to valid existing rights.

§ 4(d)—withdraws areas from public land laws, mining laws, mineral and geothermal leasing laws, and mineral material disposition, subject to valid existing rights.

§ 4(e)—authorizes measures to control fire, insects, and diseases in accordance with the Wilderness Act and H.Rept. 98-40.

§ 4(f)—requires "adequate access to such property to ensure the reasonable use and enjoyment of the property by the owner."

§ 4(g)—states that the statute is to have no effect on "the installation and maintenance of the instrumentation" if the Secretary "determines that hydrologic, meteorologic, and climatological instrumentation is appropriate to further the scientific, educational, and conservation purposes."

§ 4(h)—states that the statute is to have no effect on low-level military overflights, new units of special use airspace, or military flight training routes.

§ 4(i)—allows continued livestock grazing "where established before the date of enactment," consistent with the Wilderness Act and Appendix A of H.Rept. 101-405.

§ 4(j)(1)—authorizes "management activities to maintain or restore fish and wildlife populations and ... habitats" if consistent with applicable wilderness management plans and carried out in accordance with applicable guidelines and policies.

§ 4(j)(2)—directs that state jurisdiction over fish and wildlife is unaffected.

§ 4(k)—ensures access to "members of Indian tribes for traditional cultural and religious purposes" in accordance with the Wilderness Act and the American Indian Religious Freedom Act (P.L. 95-341), and allows temporary closures to the public of "the smallest practicable area for the minimum period of time necessary."

§ 4(l)—states that the statute creates no buffer zone.

§ 4(m)—prohibits closing "cherry-stemmed" roads (defined in the subsection) or prohibiting access on those roads, except "for purposes of significant resource protection or public safety."

§ 5—provides BLM release language for specified areas.

§ 6—designates one potential BLM wilderness area with 11,271 acres in California. Directs management as wilderness pending its designation upon the earlier of the *Federal Register* notice that the conditions incompatible with the Wilderness Act have been removed or five years after enactment. Also authorizes motorized equipment and mechanized transport for "ecological restoration (including the elimination of non-native species, removal of illegal, unused, or decommissioned roads, repair of skid tracks, and any other activities necessary to restore the natural ecosystems)," but directs use of "the minimum tool or administrative practice necessary to accomplish the ecological restoration with the least amount of adverse impact on wilderness character and resources."

P.L. 109-382—The New England Wilderness Act of 2006

Title I—New Hampshire

§ 102—designates one new USFS wilderness area and one USFS wilderness area addition with 34,500 acres in New Hampshire.

§ 104(a) & (b)—direct management in accordance with the Wilderness Act, subject to valid existing rights.

§ 104(c)—directs that state jurisdiction over and responsibilities for fish and wildlife are unaffected.

§ 104(d)—withdraws areas from public land laws, mining laws, mineral and geothermal leasing laws, subject to valid existing rights.

Title II—Vermont

§ 211—designates two new USFS wilderness areas and four USFS wilderness area additions totaling 41,652 acres in Vermont.

§ 213(a)—directs management in accordance with the Wilderness Act, subject to valid existing rights.

§ 213(b)—directs that state jurisdiction over fish and wildlife is unaffected, "including the stocking of fish ... to support the Connecticut River Atlantic Salmon Restoration Program."

§ 213(c)—authorizes "the continuance of" three specified trails and "the marking and maintenance of associated trails and trail structures."

P.L. 109-432 (Tax Relief and Health Care Act of 2006) Division C (Other Provisions), Title III—the White Pine County Conservation, Recreation, and Development Act of 2006, Subtitle B

§ 323(a)—designates 12 new BLM wilderness areas and two BLM wilderness area additions totaling 558,133 acres in Nevada.

§ 323(d)—withdraws the areas from public land laws, mining laws, mineral leasing laws, and geothermal leasing laws.

§ 324(a)—directs management in accordance with the Wilderness Act, subject to valid existing rights.

§ 324(b)—allows continued livestock grazing, "subject to ... reasonable regulations, policies, and practices" and consistent with the Wilderness Act and Appendix A of H.Rept. 101-405.

§ 324(d)(3)—directs that nothing "(A) shall constitute or be construed to constitute either an express or implied reservation by the United States of any water or water rights ... (B) shall affect any water rights in the State ... (C) shall be construed as establishing a precedent ... or (E) shall be construed as limiting, altering, modifying, or amending any of the interstate compact or equitable apportionment decree that apportions water among and between the State and other States."

§ 324(d)(4)—directs the Secretary to "follow the procedural and substantive requirements of the law of the State ... to obtain and hold any water rights."

§ 324(d)(5)(B)—prohibits funding, assisting, authorizing, or permitting for new water resource facilities within the wilderness areas.

§ 325—states that the statute creates no buffer zone.

§ 326—states that the statute is to have no effect on low-level military overflights, flight testing or evaluation, or new units of special use airspace.

§ 327—prohibits diminution of tribal rights, including access for activities, "including spiritual, cultural, and traditional food-gathering activities."

§ 328—provides BLM release language for specified areas.

§ 329(a)—directs that state jurisdiction over fish and wildlife is unaffected.

§ 329(b)—authorizes "management activities ... to maintain or restore fish and wildlife populations and the habitats" in accordance with the Wilderness Act and Appendix B of H.Rept. 101-405, "including thc occasional and temporary use of motorized vehicles if such use ... would promote healthy, viable, and more naturally distributed wildlife populations that would enhance wilderness values ... with the minimum impact necessary."

§ 329(c)—authorizes continued use of aircraft "to survey, capture, transplant, monitor, and provide water for wildlife populations, including bighorn sheep, and feral stock, feral horses, and feral burros" consistent with the Wilderness Act and Appendix B of H.Rept. 101-405.

§ 329(d)—authorizes "structures and facilities ... for wildlife water development projects, including guzzlers" if they enhance the wilderness values and "the visual impacts ... can reasonably be minimized."

§ 329(e)—authorizes, in consultation with the state, zones and periods of no hunting, fishing, or trapping "for public safety, administration, or compliance with applicable laws."

§ 329(f)—requires a cooperative agreement with the state specifying "the terms and conditions under which the State ... may use wildlife management activities in the wilderness areas."

§ 330—authorizes "such measures as may be necessary to control fire, insects, and diseases," consistent with the Wilderness Act.

§ 331—authorizes "the installation and maintenance of [hydrologic, meteorologic, or climatological] collection devices" if appropriate for "scientific, educational, and conservation purposes."

P.L. 110-229 (Consolidated Natural Resources Act of 2008)

§ 101(a)—designates one new USFS wilderness area with 106,000 acres in Washington.

§ 101(b)(1)—directs management in accordance with the Wilderness Act, subject to valid existing rights.

§ 101(b)(2)—requires consultation with interested parties to develop a trail plan "adjacent to or to provide access to" the wilderness.

§ 101(b)(3)—authorizes "helicopter access to construct and maintain a ... telecommunications repeater site ... for safety, health, and emergency services."

§ 101(b)(4)—authorizes continued use of floatplanes on one lake, subject to reasonable regulations.

§ 101(b)(5)—authorizes continued operation and maintenance of one lookout tower.

§ 101(c)(2)—requires "adequate access to private inholdings" within the wilderness.

P.L. 111-11—The Omnibus Public Land Management Act of 2009

Title I, Subtitle A—Wild Monongahela Wilderness

§ 1001(a)—designates three new USFS wilderness areas and three USFS wilderness area additions totaling 37,771 acres in West Virginia.

§ 1001(c)—directs management in accordance with the Wilderness Act, subject to valid existing rights.

§ 1001(c)—authorizes continuation of "the competitive running event permitted from 2003 through 2007 in the vicinity of the boundaries of the Dolly Sods Wilderness ... in a manner compatible with the preservation of such areas as wilderness."

§ 1001(e)—directs that state jurisdiction over and responsibilities for fish and wildlife are unaffected.

Title I, Subtitle B—Virginia Ridge and Valley Wilderness

§ 1101(a)—amends P.L. 100-326 to designate six new USFS wilderness areas and six USFS wilderness area additions totaling 39,105 acres in Virginia.

§ 1102(b)—designates one USFS wilderness study area with 3,226 acres in Virginia.

§ 1103—designates one potential USFS wilderness area addition with 349 acres in Virginia. Directs management in accordance with the Wilderness Act, but authorizes, for "ecological restoration (including the elimination of nonnative species, removal of illegal, unused, or decommissioned roads, and any other activity necessary to restore the natural ecosystems in the potential wilderness area) ... use of motorized equipment and mechanized transport" until the area is added to the designated wilderness. Adds the area the existing wilderness when "the conditions ... that are incompatible with the Wilderness

Act ... have been removed, or ... five years after the date of enactment," whichever occurs sooner.

Title I, Subtitle C—Mt. Hood Wilderness, Oregon

§ 1202(a)—designates three new USFS wilderness areas and five USFS wilderness area additions totaling 124,240 acres in Oregon.

§ 1202(c)—designates three potential USFS wilderness area additions totaling 2,770 acres in Oregon. One is to be added when "the conditions in the potential wilderness area ... are compatible with the Wilderness Act." The other areas are to be added when the specified land exchanges are completed.

§ 1202(e)—directs management in accordance with the Wilderness Act, subject to valid existing rights.

§ 1202(f)—states that the statute creates no buffer zone.

§ 1202(g)—directs that state jurisdiction over and responsibilities for fish and wildlife are unaffected.

§ 1202(h)—authorizes measures to control fire, insects, and diseases "subject to such terms and conditions as the Secretary determines to be desirable and appropriate."

§ 1202(i)—withdraws areas from public land laws, mining laws, mineral and geothermal leasing laws, and mineral material disposition.

Title I, Subtitle D—Copper Salmon Wilderness, Oregon

§ 1301(a)—amends P.L. 98-328 to designate one new USFS wilderness areas with 13,700 acres in Oregon.

Title I, Subtitle E—Cascade-Siskiyou National Monument, Oregon

§ 1405(a)—designates one new BLM wilderness area with 24,100 acres in Oregon.

§ 1405(c)(1)—directs management in accordance with the Wilderness Act, subject to valid existing rights.

§ 1405(c)(2)—authorizes measures to control fire, insects, and diseases "subject to such terms and conditions as the Secretary determines to be desirable and appropriate."

§ 1405(c)(3)—allows continued livestock grazing "if established before the date of enactment ... subject to such reasonable regulations as are considered necessary by the Secretary," in accordance with the Wilderness Act and Appendix A of H.Rept. 101-405.

§ 1405(c)(4)—directs that state jurisdiction over and responsibilities for fish and wildlife are unaffected.

Title I, Subtitle F—Owyhee Public Land Management

§ 1503(a)—designates six new BLM wilderness areas totaling 517,025 acres in Idaho. § 1503(a)(3)—provides BLM release language for certain areas.

§ 1503(b)(1)—directs management in accordance with the Wilderness Act, subject to valid existing rights.

§ 1503(b)(2)—withdraws areas from public land laws, mining laws, mineral and geothermal leasing laws, and mineral material disposition.

§ 1503(b)(3)—allows continued livestock grazing "subject to such reasonable regulations, policies, and practices as the Secretary considers necessary," consistent with the Wilderness Act and Appendix A of H.Rept. 101-405. Requires "an inventory of existing facilities and improvements associated with grazing" and authorizes "fencing around wilderness areas ... [as] appropriate to enhance wilderness values. Also authorizes the donation and termination of grazing leases or permits, in whole or in part.

§ 1503(b)(5)—requires "a trail plan that addresses hiking and equestrian trails."

§ 1503(b)(6)—authorizes "commercial services (including authorized outfitting and guide activities) ... to the extent necessary for activities that fulfill recreational or other wilderness purposes" consistent with the Wilderness Act.

§ 1503(b)(7)—requires "adequate access to the property" for private landowners within the wilderness areas, consistent with the Wilderness Act.

§ 1503(b)(8)(A)—directs that state jurisdiction over fish and wildlife is unaffected.

§ 1503(b)(8)(B)—authorizes "any management activities that are necessary to maintain or restore fish and wildlife populations and habitats" if consistent with wilderness management plans and with Appendix B of H.Rept. 101-405. Also authorizes "the occasional and temporary use of motorized vehicles, if the use, as determined by the Secretary, would promote healthy, viable, and more naturally distributed wildlife populations that would enhance wilderness values while causing the minimum impact necessary to accomplish those tasks."

§ 1503(b)(8)(C)—authorizes state use of aircraft (including helicopters) "to survey, capture, transplant, monitor, and provide water for wildlife populations," consistent with the Wilderness Act and Appendix B of H.Rept. 101-405.

§ 1503(b)(9)—authorizes measures to control fire, insects, and diseases "subject to such terms and conditions as the Secretary determines to be desirable and appropriate."

§ 1503(b)(10)—states that the statute creates no buffer zone.

§ 1503(b)(11)—states that the statute is to have no effect on low-level military overflights, flight testing and evaluation, and new units of special use airspace and flight training routes.

§ 1503(b)(12)—directs that the designation "shall not create an express or implied reservation by the United States of any water or water rights for wilderness purposes."

Title I, Subtitle G—Sabinoso Wilderness, New Mexico

§ 1602(a)—designates one new BLM wilderness area with 16,030 acres in New Mexico.

§ 1602(c)(1)—directs management in accordance with the Wilderness Act, subject to valid existing rights.

§ 1602(c)(3)—allows continued livestock grazing "if established before the date of enactment" in accordance with the Wilderness Act and Appendix A of H.Rept. 101-405.

§ 1602(c)(4)—states that the statute is to have no effect on state jurisdiction over fish and wildlife.

§ 1602(c)(5)—requires "adequate access to inholdings" and identifies a particular tract of land to be "managed as an inholding in the Sabinoso Wilderness."

§ 1602(d)—withdraws areas from public land laws except exchanges under FLPMA, mining laws, mineral materials and geothermal leasing laws.

§ 1602(e)—provides BLM release language for certain areas.

Title I, Subtitle H—Pictured Rocks National Lakeshore Wilderness

§ 1652—designates one new NPS wilderness area with 11,740 acres in Michigan.

§ 1653(a)—directs management in accordance with the Wilderness Act, subject to valid existing rights.

§ 1653(b)—authorizes continued "use of boats powered by electric motors" on two specific lakes, "subject to any applicable laws (including regulations)."

§ 1654—states that the statute is to have no effect on any treaty rights, Lake Superior water management within the national lakeshore, or "the use of motors on the surface water of Lake Superior ... [or] the beaching of motorboats at the line of demarcation" defined in § 1651(1).

Title I, Subtitle I—Oregon Badlands Wilderness

§ 1702(a)—designates one new BLM wilderness area with 29,301 acres in Oregon.

§ 1702(b)(1)—directs management in accordance with the Wilderness Act, subject to valid existing rights.

§ 1702(b)(3)—allows continued livestock grazing "if established before the date of enactment ... subject to such reasonable regulations as are considered necessary by the Secretary" in accordance with the Wilderness Act and Appendix A of H.Rept. 101-405.

§ 1702(b)(4)—requires "adequate access" to private property.

§ 1702(c)—designates as potential wilderness a specific 25-foot wide corridor, with authorized nonconforming uses allowed to continue; adds the corridor to the wilderness when the nonconforming uses have terminated.

§ 1703—provides BLM release language for specified areas.

Title I, Subtitle J—Spring Basin Wilderness, Oregon

§ 1752(a)—designates one new BLM wilderness area with 6,382 acres in Oregon.

§ 1752(b)(1)—directs management in accordance with the Wilderness Act, subject to valid existing rights.

§ 1752(b)(3)—allows continued livestock grazing "if established before the date of enactment ... subject to such reasonable regulations as are considered necessary by the Secretary" in accordance with the Wilderness Act and Appendix A of H.Rept. 101-405.

§ 1753—provides BLM release language for specified areas.

Title I, Subtitle K—Eastern Sierra and Northern San Gabriel Wilderness, California

§ 1802—designates five new USFS and BLM wilderness areas and three USFS wilderness area additions totaling 468,854 acres in California.

§ 1802(1)—states that the statute is to have no effect on "the ongoing activities of the adjacent United States Marine Corps Mountain Warfare Training Center."

§ 1803(a)—directs management in accordance with the Wilderness Act, subject to valid existing rights.

§ 1803(d)—withdraws areas from public land laws, mining laws, mineral and geothermal leasing laws, and mineral material disposition.

§ 1803(e)—authorizes measures to control fire, insects, and diseases in accordance with the Wilderness Act and H.Rept. 98-40. Directs amendment to local fire management plans for the areas. Requires "agency approval procedures ... for responding to fire emergencies" and agreements with state and local firefighting agencies.

§ 1803(f)—requires "adequate access to the property to ensure the reasonable use and enjoyment" for private landowners within the wilderness areas.

§ 1803(g)—states that the statute is to have no effect on low-level military overflights, new units of special use airspace, or military flight training routes.

§ 1803(h)—allows continued livestock grazing, consistent with the Wilderness Act and Appendix A of H.Rept. 101-405.

§ 1803(i)(1)—authorizes "management activities to maintain or restore fish and wildlife populations and ... habitats" if consistent with applicable wilderness management plans and carried out in accordance with applicable guidelines and policies.

§ 1803(i)(2)—directs that state jurisdiction over fish and wildlife is unaffected.

§ 1803(j)—states that the statute is to have no effect on horseback riding or use of recreational or commercial saddle or pack stock in the wilderness areas, in accordance with the Wilderness Act and "subject to any terms and conditions determined to be necessary by the Secretary."

§ 1803(k)—directs that outfitter and guide activities conducted under permits for the wilderness additions are to be added to any existing limits for those three wilderness areas.

§ 1804—provides BLM release language for specified areas.

Title I, Subtitle L—Riverside County Wilderness, California

§ 1851(b)(1)—designates five new USFS and BLM wilderness areas and five BLM, NPS, and USFS wilderness area additions totaling 146,242 acres in California.

§ 1851(b)(3)—states that the statute is to have no effect on "the construction, operation, or maintenance, using standard industry practices, of existing utility facilities located outside of the wilderness areas."

§ 1853(c)—designates one potential USFS wilderness area addition with 43,300 acres in California, managed "insofar as practicable as wilderness until such time as the land is designated as wilderness." Adds as wilderness upon

Federal Register notice that "(A) all uses within the potential wilderness prohibited by the Wilderness Act ... have ceased; and (B) sufficient inholdings ... have been acquired to establish a manageable wilderness unit."

§ 1853(d)(1)—directs management in accordance with the Wilderness Act, subject to valid existing rights.

§ 1853(d)(3)—withdraws areas from public land laws, mining laws, mineral and geothermal leasing laws, and mineral material disposition.

§ 1853(d)(4)—authorizes measures to control fire, insects, and diseases in accordance with the Wilderness Act and H.Rept. 98-40. Directs amendment to local fire management plans for the areas. Requires "agency approval procedures ... for responding to fire emergencies" and agreements with state and local firefighting agencies.

§ 1853(d)(5)—allows continued livestock grazing, consistent with the Wilderness Act and Appendix A of H.Rept. 101-405.

§ 1853(d)(6)—ensures access to one area "by members of an Indian tribe for traditional cultural purposes." Authorizes temporary closures to the general public "to protect the privacy of traditional cultural activities ... [of] the smallest practicable area for the minimum period necessary for such purposes."

§ 1853(d)(7)—states that the statute is to have no effect on low-level military overflights, new units of special use airspace, or military flight training routes.

Title I, Subtitle M—Sequoia and Kings Canyon National Parks Wilderness, California

§ 1902—designates one new NPS wilderness area and one NPS wilderness area addition totaling 84,656 acres in California. Also designates one potential USFS wilderness area addition with 130 acres in California, to be added to the wilderness "upon termination of the non-conforming uses."

§ 1902(1)(B)—states that the statute is to have no effect on "(i) the cabins in, and adjacent to, Mineral King Valley; or (ii) the private inholdings known as Silver City and Kaweah Han.

§ 1902(1)(C)—states that the statute is to have no effect of the potential wilderness designation on "the operation, maintenance, and repair of the small check dams and water impoundments," including the use of helicopters, on four specified lakes.

§ 1902(3)—mandates no change to the management of areas in Sequoia and Kings Canyon National Parks previously recommended or proposed as wilderness.

§ 1903(a)—directs management in accordance with the Wilderness Act, subject to valid existing rights.

§ 1903(c)—directs continued "maintenance and access to hydrologic, meteorologic, and climatological devices, facilities and associated equipment consistent with" H.Rept. 98-40.

§ 1903(e)—states that the statute is to have no effect on horseback riding or use of recreational or commercial saddle or pack stock in the wilderness areas, in accordance with the Wilderness Act and "subject to any terms and conditions determined to be necessary by the Secretary."

§ 1904—authorizes appropriations of "such sums as are necessary to carry out this subtitle."

Title I, Subtitle N—Rocky Mountain National Park Wilderness, Colorado

§ 1952(a)—designates one new NPS wilderness area with 249,339 acres in Colorado.

§ 1952(c)—adds the potential wilderness identified on the specified map upon a *Federal Register* notice that "all uses inconsistent with the Wilderness Act ... have ceased." § 1952(d)—specifies several areas explicitly excluded from the wilderness.

§ 1952(e)—directs management in accordance with the Wilderness Act, subject to valid existing rights.

§ 1952(f)—presents findings on water rights and directs that nothing in the subtitle "(A) constitutes an express or implied reservation by the United States of water or water rights for any purpose; or (B) modifies or otherwise affects any existing [federal] water or water rights."

§ 1952(g)—authorizes measures to control fire, insects, and diseases in accordance with "(1) the laws applicable to the Park and (2) Wilderness Act."

Title I, Subtitle O—Washington County, Utah

§ 1972(a)—designates 15 new BLM wilderness areas totaling 131,932 acres in Utah.

§ 1972(b)(1)—directs management in accordance with the Wilderness Act, subject to valid existing rights.

§ 1972(b)(2)—allows continued livestock grazing "where established before the date of enactment ... subject to such reasonable regulations, policies and practices as the Secretary considers necessary," and consistent with the Wilderness Act and the guidelines in Appendix A of H.Rept. 101-405 and H.Rept. 96-617.

§ 1972(b)(3)—authorizes measures to control fire, insects, and diseases in accordance with the Wilderness Act.

§ 1972(b)(4)—states that the statute creates no buffer zone.

§ 1972(b)(5)—states that the statute is to have no effect on low-level military overflights, flight testing and evaluation, or new units of special use airspace or military flight training routes.

§ 1972(b)(7)—states that the statute is to have no effect on "(A) the rights of any Indian tribe; or (B) any tribal rights regarding access to Federal land for tribal activities, including spiritual, cultural, and traditional food-gathering activities."

§ 1972(b)(8)—authorizes "the installation and maintenance hydrologic, meteorologic, and climatological collection devices ... if the Secretary determines that the facilities and access to the facilities are essential to flood warning, flood control, or water reservoir operations," in accordance with the Wilderness Act and "subject to such terms and conditions as the Secretary may prescribe."

§ 1972(b)(9)—directs that nothing "(i) shall constitute or be construed to constitute either an express or implied reservation by the United States of any water or water rights ... (ii) shall affect any water rights in the State ... (iii) shall be construed as establishing a precedent ... (iv) shall affect the interpretation ... of any other Act; or (v) shall be construed as limiting, altering, modifying, or amending any interstate compacts or equitable apportionment decrees that apportion water among and between the State and other States." Directs the Secretary to "follow the procedural and substantive requirements of the law of the State ... to obtain and hold any water rights."

§ 1972(b)(10)—directs that state jurisdiction over fish and wildlife is unaffected and authorizes "management activities to maintain or restore fish and wildlife populations (including activities to maintain and restore fish and wildlife habitats)" if consistent with applicable wilderness management plans and carried out in accordance with the Wilderness Act and applicable guidelines and policies, "including applicable policies described in Appendix B" of H.Rept. 101- 405.

§ 1972(b)(11)—authorizes "structures and facilities, including existing structures and facilities, for wildlife water development projects, including guzzlers ... if (A) the structures and facilities will, as determined by the Secretary, enhance wilderness values by promoting healthy, viable, and more naturally distributed wildlife populations; and (B) the visual impacts of the structures and facilities on the wilderness areas can reasonably be minimized."

§ 1972(b)(12)—requires a cooperative agreement with the state specifying "the terms and conditions under which wildlife management activities in the wilderness areas ... may be carried out."

§ 1972(c)—provides BLM release language for specified areas.

§ 1973—designates one new NPS wilderness area with 124,406 acres in Utah, subject to valid existing rights.

Title II, Subtitle E—Dominguez-Escalante National Conservation Area

§ 2403(a)—designates one new BLM wilderness area with 66,280 acres in Colorado. § 2403(b)—directs management in accordance with the Wilderness Act.

§ 2405(a)—withdraws areas from public land laws, mining laws, mineral and geothermal leasing laws, and mineral material disposition.

§ 2405(b)(2)—allows continued livestock grazing "if established before the date of enactment ... subject to any reasonable regulations, policies and practices that the Secretary determines to be necessary," and in accordance with the Wilderness Act and the guidelines in Appendix A of H.Rept. 101-405.

§ 2405(e)(3)—authorizes measures to control fire, insects, and diseases "subject to such terms and conditions as the Secretary determines to be desirable and appropriate" and in accordance with the Wilderness Act.

§ 2405(h)(2)—directs the Secretary to "ensure that any water rights within the Wilderness required to fulfill the purposes ... are secured in accordance with" state law and other specific provisions.

§ 2405(h)(3)—prohibits the President or "any other officer, employee, or agent of the United States ... [to] fund, assist, authorize, or issue a license or permit for the development of any new irrigation and pumping facility, reservoir, water conservation work, aqueduct, canal, ditch, pipeline, well, hydropower project, transmission, or other ancillary facility, or other water diversion, storage, or carriage structure in the Wilderness," except for new livestock watering facilities in accordance the Wilderness Act and Appendix A of H.Rept. 101-405.

§ 2405(i)—directs that state jurisdiction over fish and wildlife is unaffected. § 2405(j)—directs that management is subject to valid existing rights.

End Notes

[1] There are actually only 117 subsequent laws with wilderness designations, but the Omnibus Public Land Management Act of 2009 (P.L. 111-11) has 16 separate subtitles designating wilderness areas, most of which had been separately introduced wilderness bills and most of which have distinct management provisions. Thus, this statute is counted as 16 separate "wilderness laws" for discussing the frequency of the various statutory provisions.

[2] Congress has also established numerous other designations, including various National Park System units (e.g., national preserves and national seashores) and special management areas within areas managed by the USFS and the BLM (e.g., national recreation areas). For details on such USFS areas, see CRS Report R41285, *Congressionally Designated Special Management Areas in the National Forest System*, by Ross W. Gorte.

[3] The appendix includes a few additional substantive bills that do not designate wilderness areas, such as wilderness study bills and wilderness area management legislation, and excludes statutes with only minor boundary adjustments.

[4] See CRS Report RL31447, *Wilderness: Overview and Statistics*, by Ross W. Gorte.

[5] The Wilderness Act only designated wilderness areas within USFS lands, and thus directed management activities for the national forests by the Secretary of Agriculture. When Congress began designating wilderness within DOI agency lands in 1968, the common management directions were extended to DOI lands by the Secretary of the Interior.

[6] Many of the statutes have included provisions directing specific land exchanges, but these might or might not have involved lands within the designated wilderness areas. The only provisions discussed here are those that clearly involve lands within wilderness areas.

[7] *Winters v. United States*, 207 U.S. 564 (1908). See also CRS Report R41081, *The Wild and Scenic Rivers Act (WSRA): Protections, Federal Water Rights, and Development Restrictions*, by Cynthia Brougher for a discussion of federal reserved water rights in similarly protective legislation.

[8] See, for example, *United States v. New Mexico*, 438 U.S. 696, n.5 at 702 (1978).

[9] The Wilderness Act designated as wilderness the agency's administratively identified wilderness and wild areas—54 areas in 13 states with 9.1 million acres.

[10] P.L. 94-579; 43 U.S.C. §§1701 et seq.

[11] See CRS Report RS21917, *Bureau of Land Management (BLM) Wilderness Review Issues*, by Ross W. Gorte and Pamela Baldwin.

[12] P.L. 86-517; 16 U.S.C. §528-531.

[13] P.L. 94-588; 16 U.S.C. §§1600-1613.

[14] The first Roadless Area Review and Evaluation (RARE) was begun in 1970; it was abandoned in 1972 without recommendations, because of a challenge that the process did not comply with the National Environmental Policy Act of 1969 (NEPA, P.L. 91-190; 42 U.S.C. §§4321-4347).

[15] *California v. Bergland*, 483 F. Supp. 465 (E.D.Cal. 1980), aff'd in part, rev'd in part, 690 F.2d 753 (9th Cir. 1982). For additional information, see CRS Report RL31447, *Wilderness: Overview and Statistics*, by Ross W. Gorte.

[16] See CRS Report RS21917, *Bureau of Land Management (BLM) Wilderness Review Issues*, by Ross W. Gorte and Pamela Baldwin.

[17] States with national forest wilderness areas but not statewide wilderness bills with release language include Idaho, Illinois, Louisiana, Minnesota, Montana, South Carolina, and South Dakota. Tennessee's national forests effectively received statewide release in two

wilderness statutes (P.L. 98-578 and P.L. 99-490) designating national forest wilderness in the state.

[18] Such closure authorizations have been included in many of the statutes designating other special management areas within the national forests; see CRS Report R41285, *Congressionally Designated Special Management Areas in the National Forest System*, by Ross W. Gorte.

[19] The House Committee on Interior and Insular Affairs slightly modified the language in H.Rept. 96-617 and reproduced it in H.Rept. 96-1126 (the conference report that accompanied P.L. 96-312). However, the reference in subsequent wilderness statutes is to §108 of P.L. 96-560 that refers to H.Rept. 96-617, and thus H.Rept. 96-617 is presented here and in Appendix B.

[20] Typically when a grazing permit is relinquished, the agency reoffers it to an appropriate livestock operator.

[21] P.L. 95-341, 42 U.S.C. §1996.

[22] Statutes with only minor boundary adjustments have been excluded. A few additional substantive bills that do not designate wilderness areas, such as wilderness study bills and wilderness area management legislation, have been included.

[23] This repeats the designation of one NPS wilderness area with 25,370 acres and one potential wilderness addition with 8,003 acres in California designated in P.L. 94-544.

[24] Designated as §301(c)(1(A), §301(c)(1)(B), §301(c)(2), and §301(c)(3) in original. Should have been designated as §301(d)(1)(A), §301(d)(1)(B), §301(d)(2), and §301(d)(3), respectively.

[25] Presumably, this would still allow "species traditionally stocked before wilderness designation," even if they were exotic species, given the direction in #7 that such species "may be considered indigenous."

[26] Section 3(a)(2) transfers jurisdiction of the small amount of BLM land included in some of the wilderness designation to the USFS.

In: Wilderness Laws: Provisions, Permissions … ISBN: 978-1-61470-694-6
Editors: C. Brown, T. A. Wilson

Chapter 2

WILDERNESS: OVERVIEW AND STATISTICS*

Ross W. Gorte

SUMMARY

Congress enacted the Wilderness Act in 1964. This act created the National Wilderness Preservation System, reserved to Congress the authority to designate wilderness areas, and directed the Secretaries of Agriculture and of the Interior to review certain lands for their wilderness potential. The act also designated 54 wilderness areas with 9 million acres of federal land. Congress began expanding the Wilderness System in 1968, and today, there are 759 wilderness areas, totaling nearly 110 million acres, in 44 states. Numerous bills to designate additional areas and to expand existing ones are introduced and considered in every Congress.

The Wilderness Act defined wilderness as an area of generally undisturbed federal land, but did not establish criteria or standards to determine whether an area should be designated, because of differing perceptions of wilderness and because of the varying purposes of wilderness. In general, wilderness areas are undeveloped, and commercial activities, motorized access, and roads, structures, and facilities are generally prohibited in wilderness areas. However, in response to conflicting demands, Congress has granted both general exemptions and specific exceptions to the general standards and prohibitions.

---

* This is an edited, reformatted and augmented version of a Congressional Research Service publication, CRS Report for Congress RL31447, from www.crs.gov, dated January 24, 2011.

The federal government owns about 29% of the land in the United States, although the proportion in each state varies widely. Four federal agencies—the Bureau of Land Management, National Park Service, and Fish and Wildlife Service in the Department of the Interior and the Forest Service in the Department of Agriculture—manage most of the federal lands. These agencies manage the 110 million acres of designated wilderness, as well as many other lands. They are also protecting lands as possible additions to the Wilderness System, and are reviewing the wilderness potential of additional lands.

In total, nearly 18% of federal land administered by the four major federal land management agencies, and nearly 5% of all land in the United States, has been designated as wilderness. Alaska, because of its size and relatively pristine condition, dominates wilderness statistics—more than 52% of designated wilderness is in Alaska. In total, nearly 16% of all land (federal, state, private, and other) in Alaska has been designated as wilderness. In contrast, 3% of all land in the United States outside Alaska has been designated as wilderness.

INTRODUCTION

In 1964, the Wilderness Act[1] established a national system of congressionally designated areas to be preserved in a wilderness condition. The National Wilderness Preservation System was created with 9 million acres of Forest Service lands. Congress has since expanded the Wilderness System to more than 109 million acres (see Table 1) among some 615 million acres of land managed by the federal land management agencies—the Forest Service (FS) in the Department of Agriculture, and the National Park Service (NPS), Fish and Wildlife Service (FWS), and Bureau of Land Management (BLM) in the Department of the Interior. (See Table 2.) Federal agencies, Members of Congress, and interest groups have recommended more lands for inclusion in the System. Furthermore, at the direction of Congress, agencies have studied, or are studying, the wilderness potential of their lands. This report provides a brief history of wilderness, describes what wilderness is, identifies permitted and prohibited uses in wilderness areas, and provides data on wilderness designations as of December 31, 2010.

HISTORY OF WILDERNESS

The federal government acquired 1.8 billion acres of land through purchases, treaties, and other agreements. Initial federal policy was generally to transfer ownership to states and private ownership, but Congress has also provided for reserving certain lands for federal purposes, and over time has reserved or withdrawn increasing acreage for national parks, national forests, wildlife refuges, etc.[2] The general policy of federal lands disposal was changed to a general policy of retaining the remaining lands in the Federal Land Policy and Management Act of 1976.[3]

Beginning in 1897, management of the national forests emphasized conservation—protecting and developing the lands. It did not take long for some FS leaders to recognize the need to preserve some areas in a natural state. Acting at its own discretion, and at the behest of an employee named Aldo Leopold, the FS created the first wilderness area in the Gila National Forest (NM) in 1924. In the succeeding decades, the agency's system of wilderness, wild, and primitive areas grew to 14.6 million acres. However, in the 1950s, increasing timber harvests and recreation use of the national forests led to public concerns about the permanence of this purely administrative system. The FS had relied on its administrative authority in making these designations; there was no law guaranteeing the future of wilderness.

In response to these concerns, Congress enacted the Wilderness Act in 1964. The act defines wilderness, and prohibits or restricts certain activities in wilderness areas, while permitting other activities to occur. The act also reserves to Congress the authority to designate areas as part of the National Wilderness Preservation System.

The Wilderness System was initially endowed with the 9.1 million acres of national forest lands that had been identified administratively as wilderness areas or wild areas. The Wilderness Act also directed the Agriculture Secretary to review the agency's 5.5 million acres of primitive areas, and the Secretary of the Interior to evaluate the wilderness potential of National Park System and National Wildlife Refuge System lands. The Secretaries were to report their recommendations to the President and to Congress within 10 years (i.e., by 1974). Separate recommendations were made for each area; many areas recommended for wilderness have been designated, although some of the recommendations are still pending. FLPMA directed the Secretary of the Interior to conduct a similar review of the public lands administered by the BLM within 15 years (i.e., by 1991). The BLM submitted its recommendations to the President, and presidential recommendations have been submitted to Congress.

The 90th Congress began expanding the Wilderness System in 1968, as shown in Table 1. Five laws were enacted, creating five new wilderness areas with 792,750 acres in four states. Wilderness designations generally increased in each succeeding Congress, rising to a peak of 60.8 million acres designated during the 96th Congress (1979-1980). The largest was 56.4 million acres of wilderness designated in the Alaska National Interest Lands Conservation Act.[4]

Table 1. Additions to the National Wilderness Preservation System

Congress	Number of Laws[a]	Number of States	Number of Areas New	Acres Designated[b]
88th	1	13	54 (0)	9,125,721
89th	0	0	0 (0)	0
90th	5	4	5 (1)	792,750
91st	3	13	25 (0)	303,612
92nd	9	7	8 (1)	913,337
93rd	4	22	35 (0)	1,271,535
94th	8	23	35 (0)	2,428,327
95th	7	18	28 (5)	4,680,519
96th	7	10	71 (11)	60,753,605
97th	6	6	7 (0)	83,309
98th	21	21	177 (49)	8,530,657
99th	5	5	11 (2)	99,153
100th	7	8	22 (4)	1,422,730
101st	5	5	68 (3)	3,501,160
102nd	2	2	6 (4)	426,290
103rd	2	2	79 (14)	8,272,871
104th	2	2	1 (2)	29,970
105th	1	1	0 (1)	160
106th	8	7	18 (1)	1,081,465
107th	5	5	18 (13)	529,590
108th	2	2	15 (0)	801,784

Congress	Number of Laws[a]	Number of States	Number of Areas New (Additions)	Acres Designated[b]
109th	6	7	25 (11)	1,030,748
110th	1	1	1 (0)	106,000
111th	1	9	50 (27)	2,096,150
Total	118	44	759 (149)	109,712,833

[a] Excludes laws with minor boundary and acreage adjustments (less than 10 acres of net change).

[b] This total differs from the total of the column because of acreage revisions.

In 1977, the FS began a review (RARE II)[5] of 62 million acres of national forest roadless areas, as an acceleration of part of the land management planning process mandated by the Forest and Rangeland Renewable Resources Planning Act of 1974 and the National Forest Management Act of 1976.[6] The RARE II Final Environmental Statement was issued in January 1979, recommending more than 15 million acres (24.3% of the study area) for addition to the Wilderness System. In addition, nearly 11 million acres (17.4%) were to be studied further in the ongoing FS planning process under NFMA. The remaining 36 million acres (58.3% of the RARE II area) were to be available for other uses—such as logging, energy and mineral developments, and motorized recreation—which might be incompatible with preserving wilderness characteristics. In April 1979, President Jimmy Carter presented the recommendations to Congress with minor changes.

In 1980, the state of California successfully challenged the FS RARE II recommendations for 44 areas allocated to nonwilderness uses, with the court decision substantially upheld on appeal in 1982.[7] The Reagan Administration responded in 1983 by directing a re-evaluation of all RARE II recommendations, except in states with wilderness laws containing certain provisions known as *release language*.[8] Tensions between the Administration and Congress, and among interest groups, led to a particularly intense debate during the 98th Congress (1983-1984). A compromise version of release language, achieved in May 1984, led the 98th Congress to enact 21 wilderness laws designating 8.5 million acres of wilderness in 21 states—more laws and more acres (outside of Alaska) than any Congress since the Wilderness System was created.

Since the Wilderness Act created the National Wilderness Preservation System in 1964, Congress has enacted 117 additional laws designating new wilderness areas or adding to existing ones, as shown in Table 1. The

Wilderness System now contains 759 wilderness areas with more than 109 million acres in 44 states, managed by the four federal land management agencies, as shown in Table 3. The agencies have recommended additional lands be added to the Wilderness System; these lands are generally managed to protect their wilderness character while Congress considers adding them to the Wilderness System. Additional lands are being studied by the agencies, to determine if they should be added to the System. However, comprehensive data on the lands recommended and being reviewed for wilderness potential are not available.

Questions and discussions persist over the protection and management of areas that some believe should be designated as wilderness and others believe should be available for development. The Clinton Administration moved to protect many of the remaining national forest areas that could be designated as wilderness, initially in January 1998 with a temporary moratorium on road construction in roadless areas, and finally with a rule that prohibited most road construction and many other activities in *inventoried roadless areas* (i.e., the remaining undesignated RARE II areas). The Clinton rule was enjoined twice, and the Bush Administration promulgated a rule that has also been enjoined. The status of roadless area protection rules continues to be disputed.[9]

Questions also persist over BLM *wilderness study areas* (WSAs). These WSAs are the areas the BLM studied as potential wilderness under § 603 of FLPMA. Under § 603(b), the BLM is required to protect the wilderness characteristics of all WSAs (including areas not recommended for wilderness) "until Congress determines otherwise." Congress has designated some BLM WSAs as wilderness (included in Table 3), and generally in the same statutes, Congress has released the BLM from the requirement to protect the wilderness characteristics of certain other areas. However, release language in BLM wilderness statutes has generally been more controversial than for national forest areas.[10]

WHAT IS WILDERNESS?

The Wilderness Act defines wilderness as an area of generally undisturbed federal land. Specifically, §2(c) states:

> A wilderness, in contrast with those areas where man and his works dominate the landscape, is hereby recognized as an area where the earth and its community of life are untrammeled by man, where man himself is

> a visitor who does not remain. An area of wilderness is further defined to mean ... an area of undeveloped Federal land retaining its primeval character and influence, without permanent improvements or human habitation, which is protected and managed so as to preserve its natural conditions and which (1) generally appears to have been affected primarily by the forces of nature, with the imprint of man's work substantially unnoticeable; (2) has outstanding opportunities for solitude or a primitive and unconfined type of recreation; (3) has at least five thousand acres of land or is of sufficient size as to make practicable its preservation and use in an unimpaired condition; and (4) may also contain ecological, geological, or other features of scientific, educational, scenic, or historical value.

This definition provides some general guidelines for determining which areas should, or should not, be designated wilderness, but there are no specific criteria in the law. The phrases "untrammeled by man," "retaining its primeval character," and "man's work substantially unnoticeable" are far from precise. Even the numerical standard—5,000 acres—is not absolute; smaller areas can be designated, if they can be protected, and the smallest wilderness area—Wisconsin Islands Wilderness in the Green Bay National Wildlife Refuge—is only 2 acres.

These imprecise criteria stem in part from differing perceptions of what constitutes wilderness. To some, a "wilderness" is an area where there is absolutely no sign of human presence: no traffic can be heard (including aircraft); no roads, structures, or litter can be seen. To others, sleeping in a van or camper in a 400-site campground in Yellowstone National Park is a "wilderness experience." Complicating these differing perceptions is the wide range of ability to "get away from it all" in various settings: in a densely wooded area, "getting away" might be measured in yards, while in mountainous or desert terrain, human developments can sometimes be seen for miles.

In an attempt to accommodate contrasting views of wilderness, the Wilderness Act provided certain exemptions and delayed implementation of restrictions for wilderness areas, as will be discussed below. At times, Congress has also responded to the conflicting demands of various interest groups by allowing additional exemptions for certain uses (especially for existing activities) in particular wilderness designations. Ultimately, "wilderness areas" are whatever Congress designates as wilderness, regardless of developments or activities which some might argue conflict with the definition of wilderness.

PROHIBITED AND PERMITTED USES[11]

In general, the Wilderness Act prohibits commercial activities, motorized access, and roads, structures, and facilities in wilderness areas. Specifically, §4(c) states:

> Except as specifically provided for in this Act, and subject to existing private rights, there shall be no commercial enterprise and no permanent road within any wilderness area designated by this Act and, except as necessary to meet minimum requirements for the administration of the area for the purpose of this Act (including measures required in emergencies involving the health and safety of persons within the area), there shall be no temporary road, no use of motor vehicles, motorized equipment or motorboats, no landing of aircraft, no other form of mechanical transport, and no structure or installation within any such area.

This section prohibits most commercial resource exploitation (such as timber harvesting) and motorized entry (via cars, trucks, off-road vehicles, aircraft, or motorboats) except in emergencies. However, §4(d) provides numerous exceptions, including (a) possible continued use of motorboats and aircraft; (b) fire, insect, and disease control measures; (c) mineral prospecting conducted "in a manner compatible with the preservation of the wilderness environment;" (d) water projects; (e) continued livestock grazing; and (f) commercial recreation activities.

In addition to these exemptions, the Wilderness Act extended the mining and mineral leasing laws for wilderness areas in national forests for 20 years, through 1983. New mining claims and mineral leases were permitted for many wilderness areas, and exploration and development were authorized "subject, however, to such reasonable regulations governing ingress and egress as may be prescribed by the Secretary of Agriculture."[12] Despite this authority, no permits for on-site exploration were considered until James Watt became the Secretary of the Interior in 1981.[13] Litigation halted a drilling application in Montana that year, and Congress enacted a moratorium on wilderness area leasing and exploration in the Department of the Interior appropriations laws for FY1983 and FY1984 (P.L. 97-394 and P.L. 98-146, respectively). However, mineral rights existing on or before December 31, 1983 (or before the area was designated), remain valid, and can be developed if the right-holder chooses, under "reasonable regulations" determined by the Secretary of

Agriculture, and some mineral exploration has occurred in designated wilderness areas under such regulations.[14]

The Wilderness Act also directs that the act not alter existing federal-state relationships with respect to state water laws or state fish and wildlife responsibilities. Specifically, §4(d) (as codified at 16 U.S.C. §1133) states:

> (7) Nothing in this Act shall constitute an express or implied claim or denial on the part of the Federal Government as to exemption from State water laws.
>
> (8) Nothing in this Act shall be construed as affecting the jurisdiction or responsibilities of the several States with respect to wildlife and fish in the national forests.

However, the extent and nature of federal water rights that might arise from wilderness designations continue to be an important issue for Congress.[15]

Finally, as noted above, Congress has also enacted numerous exemptions to the Wilderness Act's standard prohibitions on activities and developments.[16] These exemptions typically apply to one or a few areas and typically authorize a particular activity (that might not be permitted under the Wilderness Act's management guidance) to be allowed to continue in the area at the level or intensity that occurred prior to the area's designation as wilderness.

DATA ON WILDERNESS DESIGNATIONS AS OF DECEMBER 31, 2010

The following tables present data on the federal lands managed by the four principal federal land management agencies and on the acreage designated as wilderness and recommended by the agencies for wilderness. The data were gathered from several agency sources, as described below. This section describes the agency land data shown in Table 2 and the wilderness data shown in Table 3.

Agency Land Data

Table 2 shows the area managed by the four major federal land management agencies.[17] The data are not directly comparable across agencies, however, because of differences in accounting practices. Differences occur in part because official ownership status often differs from managerial responsibility; for example, the FS administers 462,678 acres in Oregon that are officially BLM lands. Another complication is partial ownership, such as split estates, with the federal government owning only the surface (or the subsurface), and another owner for the subsurface (or surface) rights; similarly, some lands are managed by the federal government under easements and long-term leases, without fee simple federal ownership. Thus, the agency acreage statistics in Table 2 may overstate actual federal land ownership.

Table 2. Total Area Managed by Federal Land Management Agencies (in acres)

	USDA Forest Service	National Park Service	Fish & Wildlife Service[a]	Bureau of Land Management	4-Agency Total
Alabama	670,185	16,715	32,207	3,523	721,444
Alaska	21,956,250	52,636,434	76,624,845	74,830,122	228,190,293
Arizona	11,264,619	2,618,735	1,683,228	12,203,334	27,768,375
Arkansas	2,598,743	98,319	373,011	6,078	3,074,696
California	20,821,541	7,584,496	285,309	15,298,451	43,939,627
Colorado	14,520,965	609,880	141,135	8,346,045	23,618,868
Connecticut	24	5,719	918	0	6,661
Delaware	0	0	25,100	0	25,090
Florida	1,176,222	2,436,859	278,351	3,134	3,894,203
Georgia	867,199	39,664	481,309	0	1,380,881
Hawaii	1	357,772	298,592	0	652,258
Idaho	20,465,014	507,585	48,507	11,609,521	32,628,226
Illinois	297,713	12	87,422	0	384,543
Indiana	202,832	10,590	14,665	0	227,296
Iowa	0	2,708	69,914	0	72,132
Kansas	108,176	461	29,509	0	138,145
Kentucky	814,045	94,395	10,709	0	918,942
Louisiana	604,373	17,531	563,101	16,474	1,193,487
Maine	53,709	66,919	65,953	0	184,460
Maryland	0	40,480	45,737	548	85,793
Massachusetts	0	32,946	21,783	0	54,717

	USDA Forest Service	National Park Service	Fish & Wildlife Service[a]	Bureau of Land Management	4-Agency Total
Michigan	2,875,957	631,717	115,217	0	3,621,021
Minnesota	2,841,630	139,528	478,775	1,447	3,454,946
Mississippi	1,173,898	103,698	211,164	241	1,488,941
Missouri	1,492,596	54,382	59,960	0	1,606,371
Montana	17,082,821	1,214,184	633,744	7,967,413	26,780,287
Nebraska	352,463	5,650	173,294	6,354	537,283
Nevada	5,764,262	774,751	2,335,451	47,806,738	56,681,375
New Hampshire	735,519	8,391	22,676	0	765,961
New Jersey	0	35,362	70,262	0	105,364
New Mexico	9,417,975	376,528	326,803	13,476,982	23,580,342
New York	16,228	33,483	27,837	0	77,539
North Carolina	1,255,614	363,042	418,574	0	2,036,478
North Dakota	1,106,034	71,250	483,303	58,841	1,719,429
Ohio	241,300	19,423	8,636	0	269,059
Oklahoma	400,928	10,008	106,594	1,975	518,943
Oregon	15,687,556	192,020	566,517	16,133,793	32,558,142
Pennsylvania	513,418	49,927	9,962	0	571,817
Rhode Island	0	5	2,369	0	2,302
South Carolina	630,741	30,125	125,672	0	784,041
South Dakota	2,017,435	141,313	204,680	274,437	2,637,783
Tennessee	718,019	356,476	52,037	0	1,117,772
Texas	755,365	1,199,425	525,228	11,833	2,480,298
Utah	8,207,415	2,097,106	107,460	22,856,155	33,264,407
Vermont	399,565	8,830	33,486	0	441,357
Virginia	1,664,467	304,105	129,135	805	2,101,098
Washington	9,289,102	1,833,697	181,665	430,239	11,693,937
West Virginia	1,043,794	64,425	18,889	0	1,125,699
Wisconsin	1,533,517	61,744	199,769	2,366	1,795,554
Wyoming	9,241,610	2,344,852	70,674	18,367,513	30,023,061
Other[b]	28,581	21,768	25,217	0	2,053,564
U.S. Total	**192,909,421**	**79,706,436**	**88,906,345**	**249,714,362**	**615,060,009**

[a] Exclude marine areas.

[b] Includes the District of Columbia, American Samoa, Guam, Puerto Rico, Virgin Islands, and U.S. territories.

The data in Table 2 are from the agency sources shown below. The data are generally limited to the lands owned by the federal government (i.e., excluding easements and long-term leases) and administered by the agency or for which the agency has primary management responsibility. The following list identifies the land area data source for each of the agencies.

U.S. Dept. of Agriculture, Forest Service, *Land Areas Report (LAR)—as of Sept 30, 2010*, at http://www.fs.fed.us/land/staff/lar/LAR2010/LAR_Table_04.pdf.

U.S. Dept. of the Interior, National Park Service, *National Park Service, Listing of Acreage by State, as of 12/31/2009*; personal communication with Mike Walsh, NPS Land Resources Division, Washington, DC.

U.S. Dept. of the Interior, Fish and Wildlife Service, Division of Realty, *Annual Report of Lands Under Control of the U.S. Fish & Wildlife Service as of September 30, 2009*, Table 2—Summary by States, Associated Governments and Possessions, at http://www.fws.gov/refuges/realty/archives/pdf/2009_Annual_Report_of_Lands.pdf.

U.S. Dept. of the Interior, Bureau of Land Management, *Public Land Statistics 2009*, at http://www.blm.gov/public_land_statistics/pls09/pls1-4_09.pdf.

Wilderness Statistics

The wilderness statistics presented in Table 3 are the current acreage estimates by the agencies. Acreages are estimates, since few (if any) of the areas have been precisely surveyed. Table 3 is the acreage for areas that have been designated by Congress. In addition, the agencies have recommended areas for addition to the National Wilderness Preservation System, and continue to review the wilderness potential of other lands under their jurisdiction, both of congressionally designated wilderness study areas and under congressionally directed land management planning efforts. However, statistics on acreage in pending recommendations and being studied, particularly in the planning efforts, are unavailable.

For the FS, the same report as for general land statistics (listed above) provides wilderness data, at http://www.fs.fed.us/land/staff/ lar/ 2008/TABLE_7.htm. The BLM's *Public Land Statistics, 2009* shows designated wilderness, at http://www.blm.gov/public_land_statistics/ pls09/pls5- 4%20_09.pdf. The NPS data on designated wilderness areas are available on an NPS website, at http://wilderness.nps.gov/maplocator.cfm.

FWS designated wilderness data are in Table 8 of the agency's general lands report, listed above. The data have been updated to reflect wilderness designations enacted since the agencies' tables or lists were prepared, particularly including P.L. 111-11, the Omnibus Public Land Management Act of 2009.

As of December 31, 2009, Congress had designated 109.7 million acres of federal land in units of the National Wilderness Preservation System, as shown in Table 3. Nearly 52.5% of this land—57.5 million acres—is in Alaska, and includes most of the wilderness areas managed by the NPS (75%) and by the FWS (90%). About a third of the Wilderness System is managed by the FS, but 84% of FS wilderness area is outside Alaska. Wilderness areas have been designated in 44 states plus Puerto Rico; only Connecticut, Delaware, Iowa, Kansas, Maryland, and Rhode Island have no federal lands designated as wilderness.

Table 3. Federal Designated Wilderness Acreage, by State and by Agency (in acres and percentage of agency/federal land)

	USDA Forest Service		National Park Service		Fish and Wildlife Service		Bureau of Land Management		Total Designated Area		Share of NWPS
Alabama	41,367	6.2%	0	0.0%	0	0.0%	0	0.0%	41,367	5.7%	0.04%
Alaska	5,753,899	26.2%	33,079,611	64.8%	18,692,615	24.4%	0	0.0%	57,526,125	25.6%	52.43%
Arizona	1,345,008	11.9%	444,055	17.0%	1,343,444	79.8%	1,397,106	11.4%	4,529,613	16.3%	4.13%
Arkansas	116,578	4.5%	34,933	35.5%	2,144	0.6%	0	0.0%	153,655	5.0%	0.14%
California	5,078,011	24.4%	6,118,671	80.9%	9,172	3.2%	3,835,785	25.1%	15,041,639	34.2%	13.71%
Colorado	3,110,120	21.6%	309,805	50.8%	2,560	1.8%	205,888	2.5%	3,658,373	15.5%	3.33%
Connecticut	0	0.0%	0	0.0%	0	0.0%	0	n.r.	0	0.0%	0.00%
Delaware	0	n.r.	0	n.r.	0	0.0%	0	n.r.	0	0.0%	0.00%
Florida	74,495	6.3%	1,296,500	53.2%	51,252	18.4%	0	0.0%	1,422,247	36.5%	1.30%
Georgia	114,537	13.2%	8,840	22.3%	362,107	75.2%	0	n.r.	485,484	35.0%	0.44%
Hawaii	0	0.0%	155,509	44.0%	0	0.0%	0	n.r.	155,509	23.8%	0.14%
Idaho	3,961,864	19.4%	43,243	8.5%	0	0.0%	517,362	4.5%	4,522,469	13.9%	4.12%
Illinois	28,063	9.4%	0	0.0%	4,050	4.6%	0	0.0%	32,113	8.3%	0.03%
Indiana	12,463	6.1%	0	0.0%	0	0.0%	0	n.r.	12,463	5.5%	0.01%
Iowa	0	n.r.	0	0.0%	0	0.0%	0	0.0%	0	0.0%	0.00%

	USDA Forest Service		National Park Service		Fish and Wildlife Service		Bureau of Land Management		Total Designated Area		Share of NWPS
Kansas	0	0.0%	0	0.0%	0	0.0%	0	n.r.	0	0.0%	0.00%
Kentucky	18,132	2.2%	0	0.0%	0	0.0%	0	n.r.	18,132	2.0%	0.02%
Louisiana	8,679	1.4%	0	0.0%	8,346	1.5%	0	0.0%	17,025	1.4%	0.02%
Maine	11,233	20.9%	0	0.0%	7,392	11.2%	0	n.r.	18,625	10.0%	0.02%
Maryland	0	n.r.	0	0.0%	0	0.0%	0	n.r.	0	0.0%	0.00%
Massachusetts	0	n.r.	0	0.0%	3,244	14.9%	0	n.r.	3,244	5.9%	0.003%
Michigan	89,529	3.1%	143,758	22.8%	25,309	22.0%	0	0.0%	258,596	7.1%	0.24%
Minnesota	810,087	28.5%	0	0.0%	6,180	1.3%	0	0.0%	816,267	23.6%	0.74%
Mississippi	6,046	0.5%	4,080	3.9%	0	0.0%	0	0.0%	10,126	0.7%	0.01%
Missouri	64,119	4.3%	0	0.0%	7,730	12.9%	0	0.0%	71,849	4.5%	0.07%
Montana	3,372,525	19.7%	0	0.0%	64,535	10.2%	6,347	0.1%	3,443,407	12.8%	3.14%
Nebraska	7,794	2.2%	0	0.0%	4,635	2.7%	0	0.0%	12,429	2.3%	0.01%
Nevada	1,087,624	18.9%	309,539	40.0%	0	0.0%	2,055,005	4.3%	3,452,167	6.1%	3.15%
New Hampshire	138,418	18.8%	0	0.0%	0	0.0%	0	n.r.	138,418	18.1%	0.13%
New Jersey	0	n.r.	0	0.0%	10,341	14.7%	0	n.r.	10,341	9.8%	0.01%
New Mexico	1,387,498	14.7%	56,392	15.0%	39,908	12.2%	166,658	1.2%	1,650,456	7.0%	1.50%
New York	0	0.0%	1,363	4.1%	0	0.0%	0	n.r.	1,363	1.8%	0.001%
North Carolina	102,634	8.2%	0	0.0%	8,785	2.1%	0	n.r.	111,419	5.5%	0.10%

Table 3. (Continued).

	USDA Forest Service		National Park Service		Fish and Wildlife Service		Bureau of Land Management		Total Designated Area		Share of NWPS
North Dakota	0	0.0%	29,920	42.0%	9,732	2.0%	0	0.0%	39,652	2.3%	0.04%
Ohio	0	0.0%	0	0.0%	77	0.9%	0	n.r.	77	0.03%	0.0001%
Oklahoma	14,543	3.6%	0	0.0%	8,570	8.0%	0	0.0%	23,113	4.5%	0.02%
Oregon	2,232,231	14.2%	0	0.0%	940	0.2%	247,723	1.5%	2,476,766	7.6%	2.26%
Pennsylvania	9,002	1.8%	0	0.0%	0	0.0%	0	n.r.	9,002	1.6%	0.01%
Rhode Island	0	n.r.	0	0.0%	0	0.0%	0	n.r.	0	0.0%	0.00%
South Carolina	16,671	2.6%	15,010	49.8%	29,000	23.1%	0	n.r.	60,681	7.7%	0.06%
South Dakota	13,426	0.7%	64,144	45.4%	0	0.0%	0	0.0%	77,570	2.9%	0.07%
Tennessee	66,349	9.2%	0	0.0%	0	0.0%	0	n.r.	66,349	5.9%	0.06%
Texas	38,483	5.1%	46,850	3.9%	0	0.0%	0	n.r.	85,333	3.4%	0.08%
Utah	775,537	9.4%	124,406	5.9%	0	0.0%	260,913	1.1%	1,160,856	3.5%	1.06%
Vermont	100,870	25.2%	0	0.0%	0	0.0%	0	n.r.	100,870	22.8%	0.09%
Virginia	135,325	8.1%	79,579	25.9%	0	0.0%	0	n.r.	216,319	10.2%	0.20%
Washington	2,715,056	29.2%	1,739,763	95.0%	805	0.4%	7,140	1.7%	4,462,764	38.0%	4.07%
West Virginia	119,309	11.4%	0	0.0%	0	0.0%	0	n.r.	118,810	10.5%	0.11%
Wisconsin	46,414	3.0%	33,500	54.3%	29	0.01%	0	0.0%	79,943	4.4%	0.07%

	USDA Forest Service		National Park Service		Fish and Wildlife Service		Bureau of Land Management		Total Designated Area		Share of NWPS
Wyoming	3,111,232	33.7%	0	0.0%	0	0.0%	0	0.0%	3,111,232	10.4%	2.84%
Territories	10,000	35.0%	0	0.0%	0	0.0%	0	n.r.	10,000	13.2%	0.01%
U.S. Total	36,170,534	18.8%	44,139,471	56.5%	20,702,901	21.2%	8,699,927	3.5%	109,712,833	17.7%	100.00%
Share of NWPS		33.0%		40.2%		18.9%		7.9%			100.00%

Notes: n.r. = not relevant; the agency owns no land within the state.

End Notes

[1] P.L. 88-577. 16 U.S.C. §§ 1131-1136.

[2] See CRS Report RL34267, *Federal Land Ownership: Constitutional Authority and the History of Acquisition, Disposal, and Retention*, by Kristina Alexander and Ross W. Gorte.

[3] FLPMA; P.L. 94-579. 43 U.S.C. §§ 1701, et seq.

[4] ANILCA; P.L. 96-487.

[5] The first Roadless Area Review and Evaluation (RARE) was begun under the agency's administrative authority in 1970, but was abandoned in 1972 because of a lawsuit asserting the review had been restricted in ways that violated the National Environmental Policy Act of 1969 (NEPA, Act of January 1, 1970; P.L. 91-190; 42 U.S.C. §§ 4321-4347).

[6] Respectively: RPA; P.L. 93-378; and NFMA; P.L. 94-588. 16 U.S.C. §§ 1600-1614, et al.

[7] *California v. Bergland*, 483 F. Supp. 465 (E.D.Cal. 1980), aff'd in part, rev'd in part, 690 F.2d 753 (9th Cir. 1982).

[8] *Release language* provides direction on the timing of future wilderness review and of the management of areas not designated as wilderness until the next review. For a history of the debate over release language provisions, see out-of-print CRS Report 93-280 ENR, *Wilderness Legislation: History of Release Language, 1979-1992*, by Ross W. Gorte and Pamela Baldwin (available from the author).

[9] For more background information on and the current status of these rules, see CRS Report RL30647, *National Forest System (NFS) Roadless Area Initiatives*, by Kristina Alexander and Ross W. Gorte.

[10] See CRS Report R41610, *Wilderness: Legislation and Issues in the 112th Congress*, by Ross W. Gorte, Kristina Alexander, and Sandra L. Johnson.

[11] See CRS Report RL33827, *Wilderness Laws: Permitted and Prohibited Uses*, by Ross W. Gorte.

[12] Most lands in the National Park System and the National Wildlife Refuge System have been withdrawn from access under the mining and mineral leasing laws, while extensive BLM wilderness designations were apparently not contemplated until FLPMA was enacted in 1976. Thus, the Wilderness Act addressed mining and mineral leasing only in the national forests.

[13] Although national forests are managed by the FS in the Department of Agriculture, mining claims and mineral leases on most federal lands, including the national forests, are administered by the BLM in the Department of the Interior.

[14] Olen Paul Mathews, Amy Haak, and Kathryn Toffenetti, "Mining and Wilderness: Incompatible Uses or Justifiable Compromise?" *Environment*, v. 27 (April 1985): 12-17, 30-36.

[15] For a more thorough discussion of this issue, see out-of-print CRS Report 89-11 A, *Wilderness Areas and Federal Water Rights*, by Pamela Baldwin (available from the author of this report, Ross W. Gorte).

[16] See CRS Report RL33827, *Wilderness Laws: Permitted and Prohibited Uses*, by Ross W. Gorte.

[17] Other federal agencies, such as the Department of Defense, administer some federal land, but land and resource management is not their primary mission.

INDEX

H

I

J

L

M

N

S

T

U

V

W